THE USE OF AIRCRAFT IN AGRICULTURE

FAO Agriculture Series   No. 2
FAO Agricultural Development Paper   No. 94

THE USE OF AIRCRAFT IN AGRICULTURE

by

NORMAN B. AKESSON
and
WESLEY E. YATES

FAO Consultants

FOOD AND AGRICULTURE ORGANIZATION OF THE UNITED NATIONS
Rome 1974

First printing 1974
Second printing, with corrections 1976

P-06
ISBN 92-5-100067-0

The copyright in this book is vested in the Food and Agriculture Organization of the United Nations. The book may not be reproduced, in whole or in part, by any method or process, without written permission from the copyright holder. Applications for such permission, with a statement of the purpose and extent of the reproduction desired, should be addressed to the Director, Publications Division, Food and Agriculture Organization of the United Nations, Via delle Terme di Caracalla, Rome, Italy.

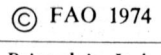 FAO 1974

Printed in Italy

FOREWORD

The use of aircraft in agriculture had its beginnings more than fifty years ago. Since that time aircraft have gradually and continuously expanded their usefulness in many phases of agricultural production, particularly in the fields of plant protection and pest control.

The world-wide advantages of aircraft are speed of operation and ability to apply material when and where ground means are either impossible or impractical.

The major objective of this bulletin is to meet the needs of agriculturists who seek information on the basic practical aspects of the use of aircraft in agriculture, which are of vital importance to agricultural production and the public welfare. These aspects of aircraft use are equally of interest to the agricultural engineer, the agronomist, the plant scientist, the entomologist, and the farmer.

The authors have included a wide range of analytical and illustrative material concerning types of aircraft, application equipment, application materials, operational analysis (including safety and maintenance), and crop uses of aircraft.

H.J. VON HÜLST
Chief, Agricultural Engineering Service
Agricultural Services Division

W.R. FURTICK
Chief, Plant Protection Service
Plant Production and Protection Division

CONTENTS

Foreword V

List of illustrations XI

List of appendix tables XVII

1. Origin of aerial application and early development 1

2. Development of an aerial application industry 9

3. Growth patterns and world levels of aerial application ... 13

4. Aerial applicator organizations 17

5. Government regulation of aerial application 19

6. Aircraft types used for aerial applications 27
    FIXED-WING AIRCRAFT 27
    ROTARY-WING AIRCRAFT 29
    GYROTATIONAL AIRCRAFT 32
    VTOL AND STOL 32
    SELECTION OF AIRCRAFT FOR THE JOB TO BE DONE 33

7. Aerial equipment for dispersing dry and liquid materials . 37
    LIQUID APPLICATION SYSTEMS 38
    Spray pumps 45
    Power sources for pumps 49

	Spray atomizers	53
	Pipe and boom size	65
	DRY MATERIAL APPLICATION SYSTEMS	68
8.	**Application techniques: Physics and technology of particle behaviour**	77
	GRANULAR SIZE DISTRIBUTION	78
	Application distribution of granulars	78
	Bait and seed spreading patterns	81
	LIQUID FORMULATIONS	84
	Particle size and effectiveness	84
	Control of spray drop size	89
	Aerial transport or drift	92
	Drift losses	96
	Spray swath patterns	98
9.	**Meteorological factors relating to aircraft application**	105
	WIND VELOCITY AND DIRECTION	106
	VERTICAL TEMPERATURE GRADIENT	106
	MICROWEATHER UNDER THE FOREST CANOPY	108
	AEROSOLS AND MICROWEATHER	108
10.	**Laboratory and field analysis of aircraft applications**	113
	SPRAY-DROP COLLECTION TECHNIQUES	113
	DRY MATERIAL COLLECTION	117
	SPRAY PATTERN DETERMINATIONS	118
	USE OF BIOASSAY FOR SWATH STUDIES	119
11.	**Operational analysis of agricultural aircraft use**	121
	ECONOMICS OF OPERATION OF VARIOUS AIRCRAFT	122
	OPERATIONAL MISSION	124
	COST OF AIRCRAFT APPLICATION	133
	CALIBRATION OF THE AIRCRAFT APPLICATOR	135

12. Flight planning, aircraft loading, and field layout 141

LOADING THE AIRCRAFT 142
FIELD LAYOUT AND APPLICATION 144
FLAGGING OR SWATH MARKING 145
FIELD MIXING AND HANDLING OF HAZARDOUS MATERIALS 148
DESIGN OF FIELD MIXING TANKS 151
HAZARDS OF PESTICIDE CHEMICALS 152
CLEANING AIRCRAFT EQUIPMENT 154

13. Aircraft flight safety and airworthiness 157

ATMOSPHERIC CONDITIONS 158
GROUND EFFECT IN TAKE-OFF AND LANDING 159
STALLS IN TURNS 160
PILOT FATIGUE 162
STRUCTURAL DESIGN OF AGRICULTURAL AIRCRAFT 163

14. Agricultural pilot training 165

15. Specific treatment practices 169

COTTON 169
RICE 170
ORCHARD AND VINE CROPS 171
USES OF AIRCRAFT FOR LARGE-SCALE PROGRAMMES 172
APPLICATION OF BIOLOGICAL MATERIALS 172
FIRE-FIGHTING AIRCRAFT 173

Appendix tables 175

Bibliographical references 209

LIST OF ILLUSTRATIONS

1. Curtiss JN-6H (Super Jenny) with 150 hp Hispano 8-cylinder engine 3
2. Dehavilland DH-48 with 420 hp 12-cylinder Liberty engine 3
3. Huff-Daland Model 5 with 150 hp Hispano engine 4
4. Aerodynamic trailing wake of a Bell 47-D4 helicopter 30
5. Aerodynamic trailing wake of a Fairchild Model 24 high-wing monoplane 32
6. Cutaway diagram of a spraying system for a small fixed-wing aircraft 39
7. Cutaway diagram of a dry materials system with a ram-air spreader on a large (600 hp) biplane 40
8. Air-driven dry materials spreader for a helicopter using electrically driven rotary feeder valves 40
9. Main control valve for aircraft sprayer, showing boom-vacuum position for positive spray shutoff 41
10. Liquid screen-filter used between pump and boom 41
11. Spray system with quick-detachable belly tank on a Piper PA-18 fixed-wing aircraft 44
12. Double air-driven pump spray system with a capacity of 3 787 litres (1 000 gals) mounted on a C-54 (Dakota) airplane 44
13. Cutaway diagram of an air-driven centrifugal pump 47
14. Operating characteristics (pressure versus flow rate) of a centrifugal pump with a discharge port 2 cm (3/4 in) in diameter.. 48
15. Cutaway view of an external-drive gear pump 50

16. Cutaway view of a roller type pump 50
17. Aircraft sprayer with centrifugal pumps driven by hydraulic motor 50
18. Schematic drawing of a hydraulic-powered aircraft showing enlarged sections of control and pressure relief valves 51
19. Cutaway view of hydraulic (oil-driven) piston type pump or motor for operating spray pumps 52
20. Spray system with electrically driven gear pump for low-volume spraying mounted on a Piper Pawnee aircraft 52
21. (a) Mechanical belt-driven centrifugal pump with electric solenoid clutch disengage mounted on a Bell 47AG helicopter 54
21. (b) Mechanical belt-driven centrifugal pump on a Stearman fixed-wing aircraft 54
22. Air-driven spray atomizer with spinning wire brush mounted on a fixed-wing aircraft 54
23. Cutaway diagram of a spinning screen (Micronair) type of air-driven atomizer 54
24. (a) Hydraulic pressure nozzles 56
24. (b) Above: Low-turbulence microfoil type nozzle for uniform large-drop sprays. Below: Solid jet nozzle recommended for large-drop-size sprays, with diaphragm-type check valve 58
24. (c) Left: Hollow cone nozzle directed with airstream for a medium-drop-size spray. Right: Hollow cone or fan nozzle directed 90 degrees toward the airstream for fine to medium sprays 59
25. Drop size versus liquid pressure for a hollow cone and a fan nozzle 60
26. Atomization (drop size) versus spray discharge direction to the airstream for a hollow cone and a fan type nozzle 61
27. Drop size versus air velocity for an 8002 fan nozzle in two positions relative to the airstream 63
28. Drop size versus nozzle orifice diameter for a hollow cone nozzle directed 135 degrees toward the airstream 63
29. Atomization (drop size) versus liquid flow rate and at pe-

XIII

	ripheral velocities shown for the spinning wire-gauze (Microfoil) devices, with data for two liquid viscosities	64
30.	Single-swath spray distribution pattern for a Piper Pawnee aircraft	66
31.	Single-swath spray distribution pattern for a Cessna 182 high-wing monoplane	66
32.	Spray boom on a Bell 47AG helicopter	67
33.	Ram-air type of dry materials spreader mounted on a Piper Pawnee aircraft	67
34.	Cutaway of a dry materials metering gate showing emergency dump gate	68
35.	Quick-loading dry materials basket system with spinning disk spreader driven by hydraulic power from the helicopter	69
36.	Bottom view of an electrically driven spinning disk type of dry materials spreader on a Hughes helicopter	69
37.	Double-spinner dry materials spreader, hydraulic-power driven, on a Stearman biplane	70
38.	Rapid bulk-handling system	71
39.	Scoop type skip loader mounted on back of service vehicle	72
40.	Unit-load bag system for dry materials	72
41.	Dry material applicators. The Razak forced-air distribution system and the Mississippi project air-power spreader system	73
42.	Cessna AgWagon spreading dry granular materials without spreader and with New Zealand type rotary cylinder spreaders on each side	74
43.	Graph of Cessna AgWagon ram-air type of dry granular materials spreader for two swath directions and at two rates of application	79
44.	Graph of Stearman aircraft with single spinning disk type of dry granular materials spreader, showing two application directions	79
45.	Graph of double spinning disk type of dry materials spreader on a Hughes helicopter	81

46. Graph of rice sowing with a Cessna AgWagon 82
47. Graphs of a Stearman aircraft with single spinning disk spreading granular fertilizer 83
48. Two ways of covering a field: progressive passes and round-robin system 83
49. Graph of theoretical deposit versus airstream (liquid drop or particle velocity) for several drop sizes and two object sizes 86
50. Graphic presentation of observed data showing logarithmic (vertical drop-size scale) versus statistical probability of drop-size distribution and the 50% or VMD (volume median diameter) line for several types of atomizers 93
51. Downwind air-transport levels of actual field-measured spray drift 94
52. Downwind air-transport and ground-level residues from several aircraft and ground-operated spray machines 95
53. Downwind drop-size recovery from field data taken with a water-base spray 96
54. Several single-swath graphs of different spray-drop-size systems used on a Stearman aircraft 99
55. Two graphs of single-swath distributions of coarse-spray and medium-spray applications from a Bell 47AG helicopter with 15 m (50 ft) boom 99
56. Graph of medium-spray application with two boom lengths on a Bell 47AG helicopter 100
57. Fine-spray distribution from a 15 m (50 ft) boom on a Bell 47AG helicopter 101
58. (a) Multiple-orifice nozzle, showing spray pattern of inverted (water in oil) emulsion 101
58. (b) Bell 47AG helicopter spraying with a narrow-range drop size microfoil, using a low-turbulence nozzle system 102
59. Closed transfer system for handling hazardous chemicals 102
60. Graph of deposited spray residue recovery from medium spray applied by a Stearman aircraft 103

61.	Field operation, showing a Hughes helicopter landing on a vehicle-mounted platform for rapid field-loading procedure	103
62.	Graph of downwind deposit versus downwind distance with medium spray under three types of weather conditions	111
63.	Plastic-basket technique for determining swath patterns, showing dry material spreading by a Stearman biplane	117
64.	Evaluation of spray swath distribution using plastic plates spaced at intervals	118
65.	Graph of spray penetration and swath distribution across a rice field, showing parts per million of active chemical caught in water-filled cups placed at rice-top level and at water level under the rice, as well as 3- and 24-hour standing water samples (below rice)	119
66.	Operation analysis (metric) for a fixed-wing aircraft at a heavy application rate	126
67.	Operation analysis (metric) for a fixed-wing aircraft at a normal application rate	126
68.	Operation analysis (metric) for a fixed-wing aircraft at a light application rate	128
69.	Operation analysis (metric) for a helicopter at a heavy application rate	128
70.	Operation analysis (metric) for a helicopter at a normal application rate	129
71.	Operation analysis (metric) for a helicopter at a light application rate	129
72.	Operation analysis (U.K.-U.S.A.) for a fixed-wing aircraft at a heavy application rate	130
73.	Operation analysis (U.K.-U.S.A.) for a fixed-wing aircraft at a normal application rate	130
74.	Operation analysis (U.K.-U.S.A.) for a fixed-wing aircraft at a light application rate	131
75.	Operation analysis (U.K.-U.S.A.) for a helicopter at a heavy application rate	131

76. Operation analysis (U.K.-U.S.A.) for a helicopter at a normal application rate 132
77. Operation analysis (U.K.-U.S.A.) for a helicopter at a light application rate 132
78. Schematic drawing of the manner of flying a field for aerial application 145
79. Schematic drawing of a liquid-spray mixing tank with mechanical paddle agitation system 151
80. Schematic drawing of a liquid-spray mixing tank with hydraulic jet system 151
81. Worker wearing rubber gloves and respirator placing a toxic chemical in the sump 153

LIST OF APPENDIX TABLES

1. World-wide aircraft use for plant seeding, protection, and nutrition as well as vector and locust control 176
2. Trends in material forms and uses of aircraft applications, areas treated, and numbers of aircraft 177
3. Some types of fixed-wing aircraft available commercially now or in the recent past 178
4. Some types of rotary-wing aircraft which are or have been available commercially 182
5. Particle density and distribution of sample aerosol dusts ... 184
6. Approximate settling rates of dry material with standard screen sizes 184
7. Terminal velocities of water drops in still air and number per given volume in relation to unit area and air volume ... 185
8. Physical properties of selected liquid materials 186
9. Twin-fluid air/liquid atomizers 187
10. Drop size and air velocity with water at 20°C (68°F) 187
11. Aircraft spray drop size range, approximate recoveries, and uses 188
12. Lifetime and fall distance at terminal velocity for water droplets 189
13. Cost of operating agricultural aircraft 190
14. Insurance schedule 191
15. Operations analysis data 192
16. Approximate costs per hectare (acre) for various aircraft applications 193

17.	Characteristic flow rates of Spraying Systems Co. hydraulic nozzles operated with water at given pressures	194
18.	Characteristic flow rates of Spraying Systems Co. hydraulic nozzles operated with water at given pressures	196
19.	Characteristic flow rates of Spraying Systems Co. hydraulic nozzles operated with water at given pressures	199
20.	(a) Acres covered for given field lengths and swath widths	200
20.	(b) Hectares covered for given field lengths and swath widths	200
21.	(a) Hectare per minute coverage for given velocities and swath widths	201
21.	(b) Acre per minute coverage for given velocities and swath widths	201
22.	(a) Required jet diameters for given tank sizes and pressures	202
22.	(b) Flow rates (water) for given jet diameters and pressures	202
23.	(a) Toxicity ratings of chemicals used in pest control	203
23.	(b) Combined tabulation of toxicity classes	204
24.	Estimated relative acute toxic hazards of pesticides to spraymen	205
25.	Conversion tables for metric and for U.K. and U.S.A. weights and measures	206

NOTE

Whereas standard FAO abbreviations for units of weight and measure are used throughout the text, the diagrams and graphs provided by the authors retain the alternative forms of abbreviation listed below:

A	=	acre
kph	=	kilometres per hour
mph	=	miles per hour
psi	=	pounds force per square inch
rpm	=	revolutions per minute

The following common abbreviations for specialized and technical terms frequently appear in this publication:

ACV	=	air-cushion vehicle
AI	=	active ingredient
EC	=	emulsifiable concentrate
ID	=	internal diameter
LV	=	low volume
µg	=	microgramme
µm	=	micron or micrometre
PTO	=	power take-off
STOL	=	short take-off and landing
ULV	=	ultra-low volume
VMD	=	volume median diameter
VTOL	=	vertical take-off and landing

1. ORIGIN OF AERIAL APPLICATION AND EARLY DEVELOPMENT

As has been true of the origins of many other mechanization developments, many "firsts" are claimed for the use of airborne vehicles to apply various plant protection and nutritional materials and to spread seeds. A documented first is a patent granted to Alfred Zimmerman, a German forester, which covers the proposed application of chemicals to control forest insects (Maan, 1967). However, other "firsts" have been brought forward — a notable one by the New Zealanders (Alexander and Tullett, 1967), who claim that one of their farmers, John Chaytor, when faced with a problem of seeding a particularly swampy and inaccessible portion of his land decided to try a tethered hot-air balloon. A sack of grass seed with a controlled discharge tube was hung from the balloon, and after the balloon was fired up, it was hauled back and forth with ropes by men stationed at each side of the field. While the authenticity of such a tale may be questioned, a more recent research project by Japanese agricultural engineers (Takanaga, 1969) investigated the potential use of unmanned vehicles, both overhead wire guided and VTOL (vertical take-off and landing) types, for applying various pesticide chemicals. The elimination of the pilot provides for an increased payload and reduces flying hazards; however, there has been no commercial acceptance of unmanned aircraft for agricultural work, with the possible exception of the use of space satellites for "remote sensing" of forest and crop areas of the world.

The use of aircraft for applying chemicals to control agricultural and forest pests was envisioned by many people in many countries prior to World War I; however, these early speculations had to deal with the realities of underpowered and understructured aircraft. Suitable aircraft and men trained in their use were not available until after World War I. This combining of aircraft, pilots, and application needs could not have occurred until the early 1920s, when the demand for cotton insect control was high in the U.S.A. and the control of forest insects, locusts, and crop pests assumed high priority in the U.S.S.R., while fertilization and seeding began to demand interest in New Zealand.

The first use of aircraft for pest control in the U.S.A. is documented in an article by C.R. Neillie and J.L. Houser in *The National Geographic Magazine* of March 1922, which told of the use of a military observation airplane, a Curtiss JN6 (Jenny), for dusting tests on catalpa trees by Dr. Houser and others at the Ohio Agricultural Experiment Station in the previous year. The lead arsenate dust used was distributed from the plane by the simple expedient of opening a bag and allowing its contents to fall through a hole in the deck of the rear or observation cockpit (Houser, 1922). This early work was done with aircraft, pilot, and crew borrowed from the U.S. Air Service at Dayton, Ohio. A proper dust hopper was later installed, but it was placed on the outside of the aircraft, while an observer had to stand in the rear cockpit and crank the agitator to aid in releasing the dust.

In early July of 1922 an outbreak of cotton-leaf worm (*Alabama argillacea*) in several of the southeastern states of the U.S.A. provided a relatively easily monitored and controlled insect which could be used to observe the effects and judge the efficacy of aerial chemical application (Coad, Johnson, and McNeil, 1922). Again the military air service was called in to provide a Curtiss H or Super Jenny (JN-6H; see Fig. 1), and first tests were run (as they had been in Ohio) by dropping the calcium arsenate dust through a hole in the deck of the rear observation cockpit. However, several hopper designs were evolved by personnel of the Delta Cotton Research Station (USDA), then under the direction of Dr. Coad, and eventually a self-feeding design with a slide-gate metering valve and a small airfoil on the hopper bottom for directing the dust away from the aircraft was found to do a satisfactory job. From this basic work in 1922 with various military aircraft including the DH-48 (Dehavilland) biplane with the famous 420 hp Liberty engine (Fig. 2), there evolved the first aerial application aircraft (1924) and Huff-Daland Models 5 and 31 (Fig. 3). Built by the Huff-Daland Mfg. Co. of Ogdensburg, New York, they were the basic aircraft of the Delta Air Service, later to become the Delta Airlines Company of passenger aviation fame.

FIGURE 1. Curtiss JN-6H (Super Jenny), with 150 hp Hispano 8-cylinder engine. Working speed 130 km/hr (81 mi/hr). Dust hopper 113 kg (250 lb), with hand-cranked dust feeder. Used on cotton in the southeastern U.S.A., 1922. *USDA photo.*

FIGURE 2. Dehavilland DH-48, with 420 hp 12-cylinder Liberty engine. Working speed 145 km/hr (90 mi/hr). Dust hopper 272 kg (600 lb), with hand-cranked dust feeder. Used on cotton in the southeastern U.S.A., 1922. *USDA photo.*

FIGURE 3. Huff-Daland Model 5, with 150 hp Hispano engine. Working speed about 145 km/hr (90 mi/hr). Dust hopper 136 kg (300 lb), with the first ram-air type spreader. *Delta Airlines photo.*

The demand for control of the boll weevil (*Anthonomus grandis*) throughout the cotton-growing south of the U.S.A., met by aerial dusting with 2.2-6.7 kg/ha (2-6 lb/acre) of calcium arsenate dust, quickly led to the development of an agricultural aircraft industry. By 1925 about 20 240 ha (50 000 acres) of cotton were dusted in the southeast, and the practice moved rapidly westward to Texas, where 20 240 ha (50 000 acres) of cotton were dusted in 1927.

While suggestions for the use of aircraft for forest insect control in the U.S.A. were sounded as early as 1918, there is no documentary evidence to show that actual forest spraying or dusting was done, either in the U.S.A. or in Canada, on anything but an experimental basis until after World War II and the discovery and use of DDT. Materials such as the arsenicals applied as dusts were neither effective nor economical for forest insect control. However, Yuill (1949) points out that aircraft were used for conducting surveys of insect damage, spotting forest fires, and aerial mapping of forest land as early as 1921, and are today widely used for evaluation of forest and range lands, as well as for general agriculture (Shay, 1970). Similarly, the potential use of aircraft for controlling mosquitoes and other disease-carrying vectors was examined in the early 1920s. But, aside from the use of Paris-green dusts for *Anopheles* larvae control (King and Bradley, 1926) and applications of various oils for control of other mosquito larvae (Ginsberg, 1931), aircraft did not enter the vector-control picture in significant numbers until after the advent of DDT.

The use of aircraft in the U.S.S.R. began with experimental applications for controlling a serious locust invasion in 1922 (Azar'yan et al., 1966). Here, spray formulations were used — which may well be a "first," since up to that time in the U.S.A. only dusts had been tried. Results on locust control looked good, but because of the small aircraft and the large spray dosages required, as well as the large areas to be treated, the work progressed slowly; one airplane could only cover 4-5 ha (10-12 acres) per hour. Dusting experiments carried out in 1924 were found to permit coverage of much larger areas. Thus, in 1926 four U-1 aircraft were able to cover about 11 000 ha (27 000 acres) with arsenical chemicals for locust control. In the U.S.S.R. experiments were also proceeding on forest insect, mosquito, and cotton insect control. Control of the nun moth (*Porthetria monacha*) was attempted in 1926, using 5.6 kg/ha (5 lb/acre) of calcium arsenate or 18 kg/ha (16 lb/acre) of sodium arsenate; only 536 ha (1 323 acres) were treated, and the results were not outstanding. Chemical application work in the U.S.S.R. really got under way with the development of the U-2 (PO-2) aircraft, capable of larger loads and greater reliability. By 1931, 65 aircraft were in use and materials for locust control were applied to 527 000 ha (1.3 million acres), for mosquito control to 2 000 ha (4 940 acres), and for control of psyllae and silkworms (*Bombycidae*) to 8 900 ha (22 000 acres).

Dusting formulations continued to be the basic ones employed until 1935, when liquid sprays were again brought into use, principally to reduce drift contamination and domestic animal deaths caused by dusts and to provide a better deposit of applied material. Since the arsenicals were never very successful when applied by air, other materials, including pyrethrum, were used and found to be a significant improvement. Dosages of 10-12 kg/ha (9-10.7 lb/acre) of pyrethrum dust were very successful, for example, in controlling pine moth larvae. But, as in other parts of the world, extensive use of aircraft for the application of pesticide chemicals began after World War II with the development of DDT and BHC (hexachlorocyclohexane).

Other New Zealanders besides farmer Chaytor took the matter of top-dressing and seeding their inaccessible steeply sloped farms very seriously, despite the negative response of the Defense Department (military authority frequently controlled all aviation at that time), which in 1926 probably reflected the general cautious approach to the use of aircraft in agriculture and forestry where no high-priority health, forest, or widespread crop disaster could be used to justify such an extreme measure. Another fourteen years were to pass before A.M. Prichard, chief pilot for the Public Works Department's Aerodrome Service, was able to muster official backing and conduct seeding tests of lupin on Ninety Mile Beach. Seed was ejected in the first tests by the simple expedient of inserting a tube 5 cm (2 in) in diameter into the sack of seed and allowing seed to fall out of the bottom of the aircraft. Angling the tube in the airstream altered the flow rate, and it was found that an 18.3 m (60 ft) swath at about 6.2 kg/ha (5.5 lb/acre) could be obtained at a delivery rate of 45.5 kg/min (100 lb/min). After this other tests were run, applying various top-dressing materials. But the point of departure for New Zealand top-dressing aviation did not occur until 1949, when D.A. Campbell of the Soil Conservation Service recognized the serious problem of trying to fertilize the impossibly hilly half of New Zealand's grassland and promoted the use of three Grumann Avengers (TBM single-engine U.S. Navy bombers) for top-dressing trials. In 1948 the Soil Conservation Service and the Royal New Zealand Air Force cooperated in installing a tank in the bomb bay of one of the TBMs, and in 1949 tests with coarse granulars, to pass through a screen with 1 cm (3/8 in) holes and to be retained on a 0.63 cm (¼ in) screen, showed remarkable spreading of the granules, even though they were simply dumped out of the bottom of the tank. Commercial acceptance was immediate. Although an analysis of top-dressing operations indicated that use of large aircraft, such as the TBM and Bristol Freighter, might be most economical and that top-dressing operations by the Air Force might continue, this did not become the case. It was soon found that commercial enterprises using smaller and more maneuverable and available aircraft were most practical, and they did, indeed, take over in the application field (Gibson, 1958). The RNZAF was persuaded to declare the Tiger Moth (Dehavilland-82) as surplus, and by the end of 1949 five operators flying eleven DH-82s plus two other aircraft were in business (Alexander and Tullett, 1967).

The DH-82 could be purchased for about $1 000 each, and after installing a simple tank and buying a loader truck, the operator was ready to go. Although the DH-82 carried a maximum load of 250 kg (550 lb), the use of local farmer strips, to which the superphosphate was hauled by truck and then spread by air over the rough New Zealand hills, proved to be a very workable and economical operation. Costs for spreading were $8.56 per ton in 1953 and decreased to $6 per ton by 1955 (1 long ton = 1 016 kg). Crosswind flying techniques were developed that produced an approximate 10 m (32½ ft) swath — which was narrow as no spreading device was used, but adequate for the load carried by the DH-82. A peak number of 182 DH-82 Tiger Moths were in use by 1956.

In the U.S.A., while early development of dusting aircraft was taking place in the southeast, the use of spray aircraft in California began in the early 1930s (French, 1947), where damage caused by highly driftable dusts was an immediate problem and the first regulations regarding residue tolerances on edible farm products had been enacted in 1927 (California Administrative Code, Title 3, Agriculture Section 2461: Injurious materials). Much later (1946) a specific ordinance aimed at controlling aerial transport or drift of calcium arsenate was put into force in California as the result of dust drift from tomatoes into nearby hay crops, subsequently killing several dairy cows (Cummins, 1946). This acted to limit use of dust, and spray soon dominated aerial application in California.

Sowing rice seed into flooded fields by aircraft was reported in California in 1929 (Bates, 1930). An Eaglerock biplane (Alexander Eaglerock A-2) with a 90 hp Curtiss OX5 engine was used, and the front cockpit was lined with canvas to make a hopper which held 227 kg (500 lb) of rice. Two spouts, or openings, were rigged in the bottom of the canvas liner, and a trip device operated by the pilot opened the spouts and discharged the seed. This method of rice seeding was immediately accepted, and although the number of aircraft used has never been large, aerial seeding, fertilizing, and pest control are almost universal practice in the rice production of California today.

2. DEVELOPMENT OF AN AERIAL APPLICATION INDUSTRY

Early uses of aircraft, in agriculture, forestry, and the like were frequently unplanned and too often were done on an emergency basis, to control unexpected epidemics of locusts or other insects when it appeared that all other means had failed. Agricultural researchers, while having been partners with the military pilots in the first experimental work, appeared to be reluctant to pursue further aircraft application research once commercial operations were begun. Thus, despite the twenty years that elapsed between the first use of aircraft for agricultural and related purposes in the 1920s and accelerated world-wide acceptance in the late 1940s, very little development of agricultural aircraft application equipment took place. It appeared as though traditional agricultural researchers as well as farm operators viewed the aircraft as temporary and all aircraft operators and pilots with considerable suspicion. It has taken many years of uneasy cooperation to reach mutual recognition and appreciation of the benefits to be gained from aircraft use in agriculture. Consequently, support for agricultural aircraft research in traditional agricultural institutions has been limited; in the U.S.A., for example, research has been accepted and conducted at only a few of the many large agricultural research universities even though the use of aircraft has reached almost every state.

The research that has formed the basis for most aircraft application equipment design and techniques of use was performed by national government groups charged with the responsibility of large-scale control programmes, such as locusts, forest insects, and vector control, particularly of mosquitoes for malaria control. The work sponsored in the U.S.A. by the Office of Scientific Research and Development during and following World War II set the capability standards of equipment and aircraft that enabled rapid transition to commercial use (Kruse, Hess, and Ludvik, 1949; Sebora *et al.*, 1946; Husman *et al.*, 1947). Even earlier government-sponsored research in the U.S.S.R. helped establish the industry there (Azar'yan *et al.*, 1966).

The aircraft used for aerial application came from World War I and later military surplus, with the exception of a few civilian utility types used during the 1930s (French, 1947). It wasn't until well into the 1950s that commercial aircraft specifically designed for application work came into being (Weick *et al.*, 1956). Even after the development of specific aircraft types there was little interest on the part of aircraft manufacturers in application equipment and the techniques of its use. A few commercial companies established in the 1930s supplied pumps, nozzles, dry-material spreaders, and the other component parts, but the development was primarily the responsibility of pilots and operators, who adapted various bits of equipment and "field engineered" their design to do the required job. As new jobs were found, application equipment was adapted, and as useful adaptations were recognized, the suppliers built to the customers' needs.

The following lists indicate the wide range of activities aircraft have been called upon to do. Some are still in the experimental stage, but all have been performed commercially. While those listed as lesser uses of aircraft are either of minor importance or emergency uses, they could become major operations.

I. Basic uses of aircraft

1. Plant protection from insects, fungi, nematodes, rodents, and birds.
2. Plant nutrition with basic plant food as well as trace nutrients.
3. Destruction of unwanted plant growth, weeds, and brush.
4. Control of plant growth, and of maturation, desiccation, and defoliation, as aids in crop harvesting.
5. Seeding of crops, grasslands and forests for food and fibre production, as well as control of water and wind erosion.
6. Locust and grasshopper control.
7. Control of insects and other vectors, as well as the survey of potential infestations.
8. Forest, crop, and rangeland fire fighting.

9. Aerial surveys, photogrammetry for examination and evaluation of fire, water, insect, and disease damage, as well as for evaluation of nutritional needs, forest production potential, and crop mapping.
10. Aids to fish production, application of disease control and nutritional materials, control of unwanted fish, plant, and algae growth.

II. Lesser uses of aircraft

1. Aerial releases of live insect predators and parasites, as well as sterile insects.
2. Application of plant pollens, release of insect pollinators.
3. Frost protection by mixing warm overhead air into colder ground-level air.
4. Protection of ripening fruit, vegetable, and nut crops by spraying with reflecting washes.
5. Aerial transport and drops of farm fencing and other building materials to remote areas.
6. Herding animals, emergency feeding, scaring off seed-eating birds, hunting livestock predators and bounty animals.
7. Rainfall and snowfall surveys, as well as reservoir, run-off, and flood-control surveys, application of evaporative control materials to reservoirs.
8. Removal of damaging rain from tree-ripening fruit by helicopter air blast.
9. Selective logging of forests.
10. Weather modification, rain-making, and storm prevention.

While no accurate estimates of the lesser uses of aircraft can be made, the extent of world-wide use in the basic categories of insect, disease, and weed control, fertilizing, and seeding has been approximately ascertained by the International Agricultural Aviation Centre. Its findings, based on questionnaires submitted to agricultural and civilian aviation authorities, are summarized in Tables 1 and 2.

3. GROWTH PATTERNS AND WORLD LEVELS OF AERIAL APPLICATION

Table 1 presents data on the agricultural use of aircraft in various countries of the world based on information from the IAAC in The Hague, as well as from a few personal sources and government statistics. Note that the unit "treated ha" is the product of hectares times number of treatments. Thus, in some of the Central American countries where cotton is sprayed 30-35 times a year, the unit "treated ha" reflects repeated treatment of the same fields. In many countries large-scale locust-control programmes, along with forest fertilization and occasional control of insect outbreaks, produce a large area figure for a given year. An example of this would be the over 50% increase in aerial spraying for control of mosquito vectors following the flooding in the San Joaquin valley of California in 1969.

Data from several countries are unavailable, both because of restrictions and a probable lack of government knowledge of precisely what amount of work is being done by aircraft. It should also be pointed out that these data are subject to rapid change. For example, in Japan, where in the recent past over 90% of the application work has been done by helicopters and as much as 80% of the applications were dusts, this trend has been reversed since 1969 because of increasing concern over dust drift and hazards to workers associated with the dusts. Also, the high cost of helicopter operation is arousing increased interest in fixed-wing aircraft. Thus it is expected that recent data for Japan will show increases in spray application by fixed-wing aircraft.

As can be seen in Table 1, the largest users of application aircraft are the U.S.S.R. and the U.S.A. with 52% and 27% of the total world area treated per year. However, it should be noted that no hard data are available for China, where it is known that Russian fixed-wing Antanov aircraft and Mikhail helicopters are used, as well as Chinese-manufactured aircraft; therefore, the total numbers and hectares treated are not available. In 1966 an estimated 4 million acres (approx. 1.6 million ha) were treated by around two hundred aircraft (Porch, 1967). From the wide range of crops and treatments used in China, plus the well-known need for locust control, it would appear logical to assume that many times the 1966 acreage is now being treated.

The trends in crops treated and formulations used (i.e., spray, dust, aerosol, and granule) can be observed in the statistics reported by various national and local governments; however, these statistics are subject to several means of reporting, so any attempt at presenting hard data to show use trends is extremely difficult. An example of a use trend is shown in Table 2, where we have attempted to show how the use of various formulations in aerial application has altered since World War II, both in the U.S.A., where aircraft have long been used, and in Yugoslavia, where aircraft use is more recent. Yugoslavia has quickly adopted aircraft use, including a significant amount of aerosol fogging and ULV with small volumes of total application per hectare, as well as larger drop sizes and spray volumes.

The use trend in the U.S.A. shows that in 1950 about 50% of the materials were applied as dusts and about 38% as sprays. Going back to 1940 would have revealed that as much as 80% of the materials were applied as dusts. This high use of dusts reflects the large demand for cotton pesticides in the U.S.A., which continued to be used in dust form until low-volume spray applications became acceptable. Note that by 1960 spray applications were ahead of dusts and by 1970 the greatly reduced spray volume (1-2 l/ha) and ULV (under 1 l/ha, or 12-24 oz/acre) had displaced the more hazardous dust formulations, particularly of such highly toxic materials as parathion.

It is interesting to note that aerial fertilizing and seeding have remained at about the same level, and no indication of a large increase in use is in sight unless large-scale forest fertilizing, now being experimented with in the U.S.A., should become as popular as it is in European countries. The increase in granular materials is the result of new translocated, or systemic, pesticides and particularly of granular soil-applied herbicides. Use of these non-drifting granular formulations is expected to increase in the near future. The use section of Table 2 shows the trends in pesticides, with significant gains in fungicides and herbicides and a consequent reduction in the percentage use of insecticides.

The total area treated in the U.S.A. nearly doubled every 10 years (the 1970 data for total use are hard statistics, but the distribution data are not) despite only a gradual increase in the total number of aircraft used. This is also reflected in the increased hours of use per aircraft, now approaching 400 hours per year, compared to the below 200 hours per year average as late as 1958 (Akesson and Yates, 1964). Purchasing new aircraft instead of using military surplus has undoubtedly had an effect on use hours. The same development has also been noted in New Zealand as new aircraft replace the DH-82 (Alexander and Tullett, 1967), which because of higher capitalization costs encourages operators to use their aircraft more hours per year.

Use data from Yugoslavia (Table 2) are very high for fertilizer materials, which are widely used on forest areas as well as on crops. Early use of aerosol fogging for forest and mosquito control is reflected in the 1962 and 1964 data, but 1969 shows a significant drop, which may reflect increased use of coarse aerosol or ULV instead of fogs. It is of interest to note that dust was never used to any extent in Yugoslavia, whereas in the U.S.A. and the U.S.S.R. much of the early work, prior to World War II, was done with dusts. It is logical that the ability to reduce total applied spray volume, as occurred following the discovery and use of synthetic pesticides, has put sprays in a favourable position relative to dusts in terms of application cost.

Forest work, fire fighting, insect and weed control, and fertilization only account for about 5% of total aircraft use in the U.S.A., but where forest fertilization work is on the increase, as in Sweden, Norway, the U.S.S.R., and Yugoslavia, this has quickly become a large part of aircraft operations. Both helicopters and fixed-wing aircraft are used in forestry fertilization as well as in other application uses. But the number of helicopters remains low — by 1973 showing only a small increase from the approximately 2% for 1960 through 1965, including the significant number of jet-engine helicopters being used for fire fighting and forest application. It is claimed that the area being covered by helicopters in the U.S.A. is 10% of the aircraft total, but this is hard to substantiate. This distribution seems to be reflected in most areas of the world, the principal exception being Japan, which until quite recently used helicopters almost exclusively for aerial application work.

Finally, the last item in Table 2 shows the total treated areas for 1960 and 1970 and the numbers of aircraft used.

A significant increase in total treated hectares is to be observed in the older areas of aircraft use, as in the U.S.A., where a 40% increase can be seen, whereas in Yugoslavia, where aerial application was more recently introduced, an increase of about 190% is to be observed. The world-wide increase is about 160%, which can be regarded as no less than phenomenal and is, of course, linked with the tremendous gains in food production made the world over through the use of pesticides and plant nutrients, as well as of newer varieties of crops which are responsive to this type of culture.

Production of foods and fibres per hectare has been increased enormously by the use of plant protection and nutrient chemicals, which aircraft have helped make possible, although they do not in themselves add to productivity. Instead, the reduced labour required to produce a given quantity of food or fibre increases the productivity of the labour force and thereby not only frees labour for other purposes, such as industry, but also permits the farm entrepreneur to direct a greater proportion of his efforts to management and reduces the physical drudgery associated with crop production by manual labour methods.

It is in this area of reducing the man-hours required for food production that aircraft can be most significant. The encouragement of aircraft use by governments and its acceptance by farmers in the past decade has marked another dramatic episode in the mechanization of world agriculture. Reduction of labour per unit of food production is therefore the primary basis for aircraft use. The ability to negotiate rough, hilly, or swampy land, to pass over irrigated or flooded fields, and to proceed at a high area rate of operation are all advantages over ground application methods. In the final analysis the ability to do the job at an economic cost fully competitive with ground equipment gives aircraft its position of advantage in food and fibre production, as well as in its many other uses the world over.

4. AERIAL APPLICATOR ORGANIZATIONS

Organizations of aerial-applicator pilots and operators have been formed in the past at various levels, in some cases primarily to fulfil a social need of simply getting together to talk about mutual problems, but most frequently these have a semiprofessional or industrial type of orientation. In addition to sociability, such organizations try to exert political influence on governmental decisions regarding regulation of the industry through licensing and operational control and act to inform their members of product development of interest to the industry.

Because of close ties to agriculture many of these organizations operating at the state level in the U.S.A. grew out of meetings and conferences conducted by the Agricultural Extension Service as a part of state university systems. In smaller countries like Japan national organization seems to have developed with no discernible local prefecture orientation, except as might be identified with delegates to the organization meetings. Most importantly, industry-sponsored groups can have a significant influence on legislation affecting themselves and their customers, as well as providing a means for disseminating information on new equipment and practices and specific data on such highly practical matters as aircraft and chemical safety.

Without these organizations agricultural aviation tends to get lost between the influential civil aeronautics agencies and the equally all-inclusive departments and bureaus of agriculture, forestry, and fisheries that exist in most countries. By way of illustration, applicators in the U.S.A. found state-level organization desirable, not only because agriculture varies greatly between states, but also because of the strong wish of local operators to set up barrier legislation that would keep out-of-state operators from invading their province. Since state regulation controlled most agricultural activities, there was little need for a national voice. National contact was made through a general aviation trades association, but the agricultural voice remained small.

Increasing regulation at the national level has taken place both in the USDA Pesticide Registration Division and through the Food and Drug Administration under the Department of Health, Education and Welfare — both now part of the Environment Protection Agency. Another national step in this direction was taken in 1966, when the FAA (Federal Aviation Agency) of the Department of Transportation was given authority to license agricultural aircraft operators as a new category at a national level. These moves toward national regulation, especially the latter, no doubt had a significant influence on the decision of agricultural aircraft operators to form the National Aerial Applicators Association (now the National Agricultural Aviation Association) in 1967. At present twenty-four of the fifty states have their own organizations and, along with an association of thirteen northwestern states, are affiliated with the National Associations.

Organization at the international level was sponsored officially in Europe in 1958 by what was first designated the European Agricultural Aviation Centre, supported by ten member countries of western Europe as an outgrowth of the OEEC (Organisation for European Economic Co-operation). By 1960 the organization had changed its name to the International Agricultural Aviation Centre, and in 1971 it had an impressive listing of twenty-five member countries in most of the world. The basic function as defined by its director, Dr. W.J. Maan (1967), is that of a coordinating secretariat and clearing-house for world-wide information on agricultural aviation. Its aims are "to promote and improve the use of aircraft in agriculture, horticulture and forestry without seeing this task as an end in itself, but as a means of encouraging and advancing the production of food and fiber for the rapidly increasing world population." Member countries and associate members (industrial companies) support the organization, which furthers its stated aims with the publication of a quarterly journal, *Agricultural Aviation*, and has also sponsored four international agricultural aviation meetings and published most informative proceedings of the papers presented at each of the meetings. A publication on agricultural aviation was authored by Dr. Maan and published by FAO in 1965. Two editions of the *Handbook for agricultural pilots* have been published by IAAC.

5. GOVERNMENT REGULATION OF AERIAL APPLICATION

Aircraft using the airspace of any country for any purpose must be registered in that country, usually with a civil aviation body having the power to require inspection and compliance with regulations governing aircraft structure and engine maintenance and to issue certificates of airworthiness. In addition, such agencies are authorized to issue various levels of pilot licences based on competence and physical-fitness examinations, as well as a prescribed number of hours of piloting or flying time.

Although in the past this has usually been the limit of the civil aviation agency's jurisdiction in agricultural aviation, such agencies are more frequently expanding their authority as increased regulations are being applied on agricultural aircraft. As noted earlier, in the U.S.A. the Federal Aviation Agency now licenses agricultural pilots (since 1966) as a specific category, and has thus extended its influence further into the agricultural aviation sector. This new licensing category is authorized under part 137, title 14, chapter 1-G, Agricultural Aircraft Operations, of the Federal Aviation Regulations. Since this is the first time that the FAA has entered this sector, it might be instructive to examine some of the salient points of its structure. More details concerning U.S. regulations and those of several selected countries are to be found in the WHO survey *Control of pesticides: a survey of existing legislation* (1970), as well as other publications on pesticide regulations published by the same organization.

The basic enabling legislation legalizing the addition of part 137 to FAA regulations is contained in section 3077 of the Civil Aeronautics Act of 1958, which states that it is the intent of Congress to authorize the FAA to place such restrictions upon aircraft engaged in dusting and spraying as are necessary for the protection of persons and property on the ground. In addition, section 607 provides for the licensing of agricultural aircraft operators.

One of the reasons given for the need for further regulation of agricultural operations was the inadequacy of the certificate of waiver, which had been granted to agricultural operators as a matter of course (helicopters did not require this) so as to enable them to work under the prescribed minimum 152 m (500 ft) flying height required of all civil aircraft. In fact, this is not a new concept, and it is used in other countries, including Canada and the U.K., at this time; however, it is a negative control, and with the increasing use and hazards of using agricultural aircraft, the FAA felt obliged to put the new part 137 into effect. Two basic definitions established in part 137 are pertinent to any discussion of agricultural aviation (Barnes, 1953) and will be used in this text. These are as follows:

137.3 *Agricultural Aviation* means the operation of an aircraft for the purpose of (1) dispensing an economic poison, (2) dispensing any other substance intended for plant nourishment, soil treatment, propagation of plant life, or pest control, or (3) for engaging in other activities directly affecting agriculture, horticulture, or forest preservation; and

Economic Poison means (1) any substance or mixture of substances introduced for preventing, destroying, repelling or mitigating any insects, rodents, nematodes, fungi, weeds and other forms of plant and animal life or viruses, except viruses on or in living man or other animals, which the Secretary of Agriculture shall declare to be a pest, and (2) any substance or mixture of substances intended for use as a plant regulator defoliant or desiccant.

Two grades of licence are provided for under the licensing provision:

1. *Private operator-pilot.* The applicant must hold a current U.S. private, commercial, or air-transport pilot certificate and be properly rated for the aircraft to be used. (This does not permit operation for hire or other than over his own property.)

2. *Commercial operator-pilot.* The applicant must have the available services of at least one person who holds a current U.S. commercial or air-transport pilot certificate and is properly rated for the aircraft to be used. The applicant himself may be the person available.

The licensing regulations require tests of the following knowledge and skills:

1. Steps to be taken before starting operations, including a survey of the area to be worked.
2. Safe handling of economic poisons and proper disposal of the used containers.
3. General effects of economic poisons and agricultural chemicals on plants, animals, and persons, with emphasis on those normally used in the areas of intended operations; precautions to be observed in using poisons and chemicals.
4. Primary symptoms of poisoning of persons from economic poisons, the appropriate emergency measures to be taken, and the location of poison control centres.
5. Performance capabilities and operating limitations of the aircraft to be used.
6. Safe flight and application procedures.
7. Skill in the following manoeuvres, to be shown in any of the aircraft (of agricultural configuration) specified at that aircraft's maximum certified take-off weight or at the maximum weight established for the special-purpose load, whichever is the greater:

   (a) Short field and soft field take-off (airplanes and gyroplanes only).
   (b) Approaches to the working area.
   (c) Flare-out.
   (d) Swath runs.
   (e) Pull-ups and turnarounds.
   (f) Rapid deceleration or quick stop (in helicopters only).

Licensing under part 137 does not cover precisely what materials will be discharged from the aircraft, nor does it specify the types of application equipment or how these are to be used. "Congested area" limitations still apply and are in essence similar to the previous issuances of a waiver. Now flights over congested areas require written approval from the appropriate authority and public notice prior to the performance of the flight. Subpart D specifies that records be kept of all agricultural operations performed and held for one year, having them available for inspection upon request.

In addition to the FAA, three U.S. Federal government departments were formerly involved: USDA (Agriculture), HEW (Health, Education and Welfare), and Interior, because of its concern with fish and wildlife.

The basic enabling legislation for the USDA functions was the Federal Insecticide, Fungicide and Rodenticide Act of 1947 with the amendments of 1959, which broadened its coverage. All pesticides, biologicals, and other materials must be submitted for registration, along with suggested use, dosage and manner of application, and caution against any application hazards. Substantiating data must be supplied by the applicant as to the efficacy of the material when used as prescribed, which may be obtained from tests conducted by (*a*) the manufacturer himself, (*b*) Federal agency personnel, and (*c*) Agricultural Experiment Station workers in the state universities or other recognized authorities.

In 1971 the various functions in the several departments noted were all brought under the new Environmental Protection Agency (EPA), which now has control of all potential pollutants, including pesticide chemicals. In 1972, the new Federal Environmental Act went into effect, ultimately replacing the 1947 FIFRA.

Because pesticide chemicals constitute a potential health hazard, they are regulated under the Miller Amendment (1954) to the Federal Food, Drug, and Cosmetic Act. Administered by the Food and Drug Administration (FDA), it provides the basic tolerance limits or acceptable levels for specific chemicals on various produce, including animal products, marketed in the U.S.A. The basic information for the tolerance limits is furnished by the manufacturer of the product, with the various authoritative sources as noted, including those who may be able to furnish information on the effects of the material through feeding tests on various animals.

The final hurdle in the registration of a chemical material is its chemical impact on fish and wildlife where these may be contacted directly (as in forest or range spraying) or indirectly (as may arise when agricultural uses cause contamination of air, water, and soil). Test data are required from reliable investigations into the potential damage or hazard to wildlife, and specific tests are mandatory in the case of fish and molluscs, where the "food chain" sequence may become operative if the chemical is not easily degradable. Formerly, after clearing the Department of the Interior the information was returned to the USDA for registration if all departments agreed.

State-level regulations have generally accepted Federal standards on tolerances and labeling of chemical uses. In California, as in most states, laws passed in 1960 adopted Federal standards and brought state regulations into conformity with Federal laws. States may, however, have additional regulations more specifically governing application and use of agricultural chemicals. California, for example, regulates agricultural chemicals under its Agricultural Code, which contains two provisions for the intensification of regulation in California, aside from the normal requirement of registration of all materials used for pest control. The first of these is found in the *Restricted Herbicides* section, which includes all the various weed killers of the phenoxy type and a few others, and the second is the *Restricted Material* section, which includes many insecticides, such as arsenicals and certain carbamates (Temik) and phosphates (parathion, TEPP, bidrin). No specific aircraft equipment is required for applying the restricted materials of either category. Control of the application of all materials rests at the county level, and for either restricted herbicides or restricted materials a permit must be obtained, specifying where, when, and in what quantity the chemical is to be used. This permit will not be given if the county official sees a possible danger to crops, people, animals, or wildlife.

A further control, designated *Hazardous Area*, applies specifically to herbicides which can cause significant drift damage to susceptible crops. In this case, application of the phenoxy herbicides, as well as certain others (propanil and dicamba), is not permitted in specifically designated areas primarily associated with susceptible crops (such as grapes, cotton, and prunes) during the growing season of these crops — in California from March 15 to October 15. Approval by county officials of the type of spray nozzle and of the operating pressure and placement of nozzles on the aircraft, as well as of the weather conditions desirable for minimizing drift, is required for all uses of injurious herbicides in or out of hazardous areas. But in hazardous areas more specific regulations apply. Only a jet-spray nozzle without a whirl device for producing a fan or other dispersion of the spray can be used; the nozzle must have an orifice no smaller than 16 mm (0.63 in) in diameter and be operated at a pressure no greater than 3.2 kg/cm^2 (45 lbf/in^2), directed with the slip stream or not more than 10 degrees in a downward direction. Helicopters flying at less than 88 km/hr (55 mi/hr) can direct the nozzles 90 degrees downward to the slip stream, but not forward into the slip stream.

Other restrictions regarding weather conditions in hazardous areas prohibit discharge of spray above 3 m (10 ft) and spraying in high winds above 8 km/hr (5 mi/hr), and a few areas also require that spraying must stop when temperatures reach a prescribed limit. Records of all pertinent meteorological and application data are required and must be submitted to the county official within forty-eight hours of application completion.

The general aircraft sprayer regulations (applicable also to hazardous areas) prescribe shut-off valves for each nozzle or check valves (one-way) and a boom pressure relief (suck-back) system (Fig. 9) to reduce leaking of boom and nozzle in turns or while ferrying. No aerosols (construed to mean an average particle size of under 25 microns) are permitted.

With increasing pressure to reduce the use of slowly degradable, or hard, pesticides, principally chlorinated hydrocarbons, the restricted materials category has been extended in the California regulations. Such pesticides as dieldrin, endrin, heptachlor, and toxaphene, plus certain avicides, are included in the restricted materials category and are subject to other regulations generally limiting and reducing their use. Most recently, mercurial fungicides have been removed entirely from registration in California, as well as by Federal regulation, thus stopping their use completely.

As noted in the WHO publication *Control of pesticides* (1970), Czechoslovakia and the U.S.S.R. have specific regulations concerning aircraft application equipment, as well as meteorological conditions that limit aircraft use. Several countries exempt granular forms of pesticides from as stringent controls as those applied to sprays.

Most countries have some form of tolerance or permitted residue levels in food and feeds as well as animal products. Many exporters of agricultural products have found their produce barred from entry if it exceeded the tolerances set for these products in the importing country.

It is quite evident that regulation of pesticide use has gained the attention of people and governments the world over. To lend assistance to legislative formulation and reduce conflicting regulations in various countries, WHO has attempted to evaluate the problems to be overcome and the limitations to be faced in order to maintain the benefits and yet reduce the potential hazards of pesticides (e.g., Barnes, 1953; WHO *Specifications for pesticides used in public health*, 1967). The FAO *Guidelines for legislation for sale and marketing of pesticides* was published in 1969.

The increasing restrictions being placed on agricultural aircraft reflect the pesticide regulation controversy and its close relation to aircraft as one of the most widely used and readily acceptable means of applying pesticides. Aircraft operators, feeling that their future is at stake, are siding with chemical manufacturers in their attempt to stem the flow of additional regulations. It should be rather clear, however, that more regulations of a specific nature are likely to be applied to agricultural aviation in two basic sectors. The first will be through the regulation of all pesticide use, such as proposals to regulate pesticides on the basis of their toxicity or hazard to crops, man, and animals, and to categorize them into such classes as "general," "restricted," and by "special permit only." One conclusion to be drawn from such proposals is that many pesticides will fall into the last two classes, so better training and control of the people who advise farmers on the proper materials to use and sell them their particular products will be necessary. Licensing of pesticide advisors and sales people has already been instituted in California, where chemical advisors and salesmen are now required to register and be licensed before being permitted to prescribe chemical use.

The second area of necessary regulation regards the type of equipment, basically the degree of spray atomization for drop-size control and the chemical formulation, such as the use of granular materials for minimum drift. It is interesting to note that in many countries application regulations contain a generalized statement to the effect that application equipment shall be of an acceptable type and used in such a way as to reduce to a minimum any contamination of adjoining crops and habitation. Unfortunately, present practice in much of the world — particularly with reference to large-scale, fine-spray, low-volume treatments — tends to assume that losses from treated fields at the time of application should be accepted by others in the immediate area as a minimum price to pay for the common good being done. This concept has never found favour with the property laws of most countries. Property ownership and trespass laws are fundamental precepts and are the basis for lawsuits involving public liability and property damage caused by misapplied pesticides (*Crop dusting: legal problems in a new industry*, 1953).

Increased regulation of all pesticide applications, both ground and aerial, will take place in the future, and the best that can be done will be to try to guide and influence this regulation so as to make possible continued aircraft development within whatever limitations the laws may prescribe. In the U.S.A. the EPA has the responsibility for implementing the Federal Environmental Pesticide Control Act, which over a period of four years will organize, nationwide, the entire control programme of pesticide manufacture, registration, and use. Standards for licensing applicators and regulations governing equipment and techniques of its use are also to be included. In addition, under the Occupational Safety and Health Act (OSHA) work has begun on new regulations governing worker safety and uniform practices for pesticide-application machinery, as well as standards for worker reentry to treated fields.

6. AIRCRAFT TYPES USED FOR AERIAL APPLICATIONS

FIXED-WING AIRCRAFT

Table 3 shows a range of fixed-wing aircraft in common use. Since the list is by no means all-inclusive, it is probable that some good types have been omitted. Also, no attempt was made to include military or ex-military aircraft, although these are frequently called up, especially in emergency situations, such as vector control and other similar large-scale one-of-a-kind operations. In the U.S.A. ex-military aircraft in agricultural work are probably represented in largest numbers by the PT-17 Boeing Kaydet or Stearman biplane and Navy N3N-3 primary trainers. In all, 10 346 Kaydets and 896 N3N-3s were built. It is estimated that about two thousand of these, largely rebuilt and fitted with R-985 Pratt & Whitney Wasp Jr. 450 hp engines or with R-1340 P&W Wasp 600 hp engines, are still in use. How many more are still housed in aerial applicators' hangars is difficult to say, but the obvious limitation in the use of these aircraft relates to the shortage of engines, not of basic frames and wings. The P&W Wasp Jr. 450 hp engine was originally used on the Vultee BT-130 Valiant, not on the Kaydet. The 6 407 BT-130s built were little used in agriculture or forestry except as a source of engines for the PT-17. Limited production of the P&W R-985 engine was undertaken in Canada in the early 1950s, and the Dehavilland Beaver was fitted with these as rebuilt engines.

Requirements for agricultural aircraft do not differ too greatly from those of a good general-utility airplane, and the basic characteristics have been laid down by several designers. Probably the best known is Weick, who designed the AG I and II as experimental planes and later the Piper Pawnee series. The Piper Pawnee is probably the most widely used agricultural aircraft in the world today, although hard data on the numbers of Antanov (U.S.S.R.) aircraft in use are not available, and could exceed the Piper line.

It is interesting to note that Weick's Pawnees are low-wing types, while the Antanov and the Grumman Ag Cat are both biplanes. Most agricultural aircraft are equipped with the "tail landing" type gear, with the notable exceptions of the New Zealand Fletcher FU-24 and the Australian Airtruk, which have tricycle landing gear, probably in consideration of crosswind landing requirements.

The variety of wing designs used alter the loading and flight speeds as well as the turn characteristics of the aircraft. The Cessna Ag series, for example, uses a wing section that permits somewhat higher maximum operating speeds of around 200 km/hr, compared to the 160-180 km/hr (100-110 mi/hr) maximum of the Pawnee and Commander line.

While the aircraft used in agricultural work may include a wide variety of configurations, including single- or multiple-engine, high-wing or low-wing, monoplane or biplane types, the most widely used aircraft follow the design criteria established by Weick and others in the early 1950s (see Weick, 1956: Gibson, 1958):

1. They are capable of carrying a high payload to low gross weight ratio, which requires high-performance engines, clean aerodynamics, and a minimum of extra loading in fuel or instrumentation.
2. They must be capable of taking off (to a height of 16 m, or 50 ft) and of landing on unimproved strips, with no more than a 400 m (1 300 ft) take-off distance at sea level.
3. They should cruise at around 160 km/hr (100 mi/hr), and whether or not they are provided with wing flaps and slots, their stall speed should be low (65-100 km/hr, or 40-60 mi/hr) for landing on unimproved strips (in which case large tires are also desirable).
4. They have good stability and controllability, especially in turns, and control pressures should be light but responsive in order to reduce pilot fatigue.
5. They provide for unrestricted forward and downward visibility, as well as side vision insofar as possible.
6. For crash protection, which must be considered essential, engine and load should be placed forward of the cockpit, and a crash structure specifically designed to bend outward, as well as a roll-over structure to protect the pilot, may be provided.
7. Other safety features, which should all be mandatory, are simple

flight and dispersal controls with easily distinguished manual identification of each, no equipment protrusions or sharp levers in the cockpit, and a shoulder recoil harness with securely fastened anchor bolts.
8. Dust- and vapour-proof closed cockpits are increasingly desirable, and pressurized and air-conditioned cockpits are now available.
9. Liquids are loaded from the bottom of the tanks (Figure 6); dry materials have to be loaded from the top, but pilot protection is provided by large, dust-tight top doors for easy filling and also for cleaning and repairing of the hopper.
10. The design facilitates inspection of aircraft structure, control linkage, engine and dispersal equipment, as well as washing and cleaning of the entire plane.
11. Simple construction is used for ready maintenance by aircraft mechanics, and corrosion-resistant materials are employed in the aircraft structure and dispersing equipment.

Most of the aircraft shown in Table 3 have the basic configuration and characteristics of the above general list of specifications. Horsepower listings are the maximum at take-off and average 300 to 600 hp for aircraft with a gross weight averaging between 1 600 and 2 000 kg (3 500-4 400 lb) and with a payload under the restricted agricultural category of 600-1 000 kg (1 320-2 200 1b). The most commonly used fixed-wing aircraft are relatively small in size with a high engine-power rating and are rugged in design and structure for use on rough landing strips. While improved hard-top strips would be highly desirable, this is not likely to be the rule in agricultural flying.

ROTARY-WING AIRCRAFT

Table 4 indicates some aircraft of the rotary-wing or helicopter type which are available and have been used for agricultural work. Again, not all the machines that have been made or used appear, and present military as well as ex-military types are not listed. As in the case of fixed-wing aircraft, many military types are used particularly in emergency conditions, such as to disperse vector-control chemicals.

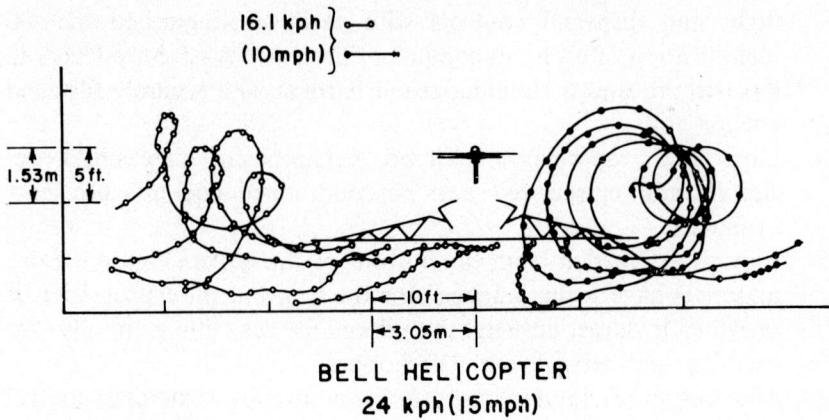

FIGURE 4. Aerodynamic trailing wake of a Bell 47-D4 helicopter. Cine camera used to track hydrogen-filled balloons. Speed 24 km/hr (15 mi/hr).

Rotor configuration may consist of one or two rotor units, either fore and aft, or side by side and synchronized to interweave. Rotors may have two, three, or four blades each. Single-rotor types usually have a vertical tail rotor to counter the rotation of the main blades. Motion of the helicopter in any direction is obtained by "flapping" or feathering the rotor blades once each revolution. The blades then can be made to "tilt" toward the desired direction, thus drawing the fuselage toward the direction of the blade tilt.

Cyclic pitch is built into all rotors (with the exception of the non-powered rotors of autogyros), which means that the blade advancing into the direction of the helicopter motion is pitched less than when on the opposite side, or moving with the direction of motion. Thus the effect of this cycling makes the air downwash appear different on the two sides of the advancing helicopter (Fig. 4; Akesson, Yates, and Burgoyne, 1966; Wooley, 1963). As the helicopter moves from hovering to increasing forward speed, the rotor downwash changes from a closed torroid to the horseshoe vortex pattern characteristic of the fixed-wing aircraft. Thus the downwash is maximum vertically when the aircraft is hovering, but the air is directed back at an increasing angle as forward velocity is increased. While the total energy in the rotor wake increases with forward speed, the downwash effect on a given ground area under the helicopter is rapidly reduced; at helicopter flight speeds of 80-95 km/hr (50-60 mi/hr) the downwash resembles that of a similar size (power and weight) fixed-wing airplane (Fig. 5). As can be seen in Tables 3 and 4, the helicopter does not have the payload capacities for a given horsepower that the fixed-wing airplane does; however, the ability to take off from and land in practically any spot large enough to clear the rotor (including platforms on top of service trucks), as well as to make shorter turns and to manoeuvre around rough terrain while applying chemicals, frequently makes up for the low payload. It is also to be noted that up to a 15 m (50 ft) spray boom has been used on a helicopter with an 11 m (37 ft) rotor, and that the 20-30 m (65-100 ft) swath obtained makes for high productivity. The greatest deterrent to helicopter use has been the high initial cost and maintenance required of the cyclic controlled rotor, which consists of closely machined and carefully balanced, highly loaded rotating elements. In recent years this cost has been reduced somewhat, but still accounts for a considerable cost difference between fixed- and rotary-wing aircraft.

Extra flying skills are required of the helicopter pilot, particularly when operating at low speeds close to the ground. The ability to hover and move slowly increases the downward force of the air, and the often repeated "current of tremendous downwash" from the helicopter is not in fact available unless the machine is flown at a low speed of 25-33 km/hr (15-20 mi/hr), when the rotor forces are directed almost entirely downward.

The patterns of air flow from fixed-wing and rotary-wing aircraft are illustrated in Figures 4 and 5 (Akesson and Yates, 1963). Gravitationally neutral balloons (filled with hydrogen) were placed in cages on a boom fitted in the manner of a spray boom beneath the aircraft wing on the helicopter rotor. On a given course and at a precise point the balloons were released from the cages and followed the wake of the air displaced by the aircraft. Cine cameras followed the balloons as they were entrained by the air wake. The graphs shown were made by plotting the balloons from the film. As can be seen, the displacement by the rather comparable-sized helicopter and fixed-wing airplane was remarkably similar. The air is displaced outward in each case, but at a low velocity. Note the scale indicating distance between balloons, where greater distance indicates higher velocity. The highest velocity occurs at the wing and rotor tips, where wing-tip vortices are frequently observed when an aircraft is applying sprays or dusts.

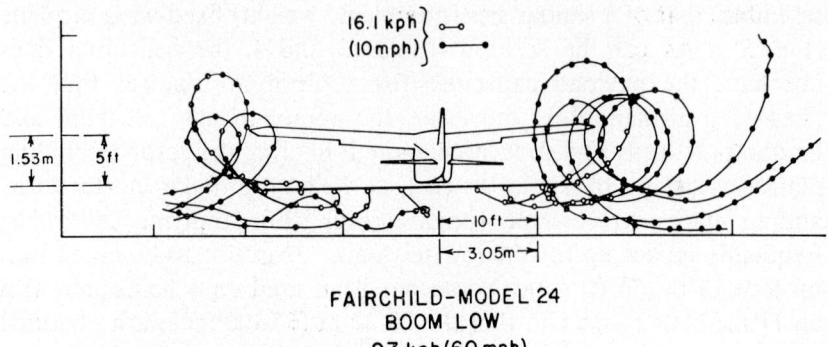

FAIRCHILD-MODEL 24
BOOM LOW
97 kph (60 mph)

FIGURE 5. Aerodynamic trailing wake of a Fairchild Model 24 high-wing monoplane. Gravitationally balanced (hydrogen-filled) balloons trace air wake. Boom 1.53 m (5 ft) below wing; speed 97 km/hr (60 mi/hr).

Gyrotational Aircraft

The autogyro is a special type of rotary-wing aircraft with a nonpowered rotor. Developed before helicopters became practical, these aircraft are capable of STOL (short take-off and almost vertical landing) due to the rotating wing or rotor, which maintains aerodynamic lift, like the helicopter, even when the aircraft is moving slowly. The propulsion engine is of a standard propeller type, and the forward motion of the aircraft causes the rotor to autorotate, thus providing the lift for the plane. A drive from the propulsion engine to the rotor to "spin it up" before take-off makes possible the "jump," or very short take-off, capability of the autogyro. These aircraft were investigated as early as 1929 for possible pesticide application use. More recently, very small (90 hp) gyrocopters called agrocopters (Benson, U.S.A., and Campbell, U.K.) have been suggested as possible carriers for mosquito aerosoling equipment.

VTOL and STOL (Vertical and Short Take-off and Landing Aircraft)

This designation could include helicopters inasmuch as they take off and land vertically without forward motion; however, the name is generally used to cover all of the many propeller and fan (ducted or other) types of machines that depend on relatively small rotors, aimed downward, for their vertical motion. This small diameter generally indicates a high loading (kg/cm^2) of the rotors, which, in turn, causes lowered rotor efficiency and high power requirements unless operated within specific limitations. By placing a shroud or duct around the fan at least a part of this lowered efficiency is regained. At present, the ACV (air-cushion vehicle) is the only practical version of this aircraft. The limitations of this aircraft arise from the need for developing an air cushion between the aircraft rotor or propeller and ground or water surface. The vehicle cannot go out of this ground effect (maximum height about 10% of its diameter, or 0.3 m for a machine 3 m in diameter) created by the enclosed air discharge or curtain. This, of course, severely limits the machine and generally means it is not suitable for agricultural work, except possibly on waterways. A separate propulsion fan is usually used to move the machine on the air cushion provided for by the main engine and fan (Sweeny and Nixon, 1968).

In order to move out of the ground effect of the air cushion the VTOL must offer (*a*) the necessary power to lift a reasonable load, thus requiring high-performance engines; (*b*) controls that permit stabilizing the machine and manoeuvring it out of the "stabilizing" ground effect; and (*c*) greater efficiency, lower costs or some other persuasive reason why it should be used instead of a helicopter. Thus far, these criteria have not been met, so no practical aircraft of this type is in use, although investigators report hope for the future of such machines (Muller, 1969). Many other small ACV machines have been built and used primarily by individuals or hobbyists, although there is a potential use for commercially built ACVs in mosquito control, and the English Hoveraire Ltd. has proposed this type of craft for spray applications.

STOL aircraft obtain relatively short take-off and landing characteristics by means of high lift at low air speed, as well as from wing flaps and slots; however, such wing designs sacrifice air speed, and this, plus added costs, reduces STOL use.

SELECTION OF AIRCRAFT FOR THE JOB TO BE DONE

As noted earlier, the survey made by aircraft operations specialists regarding the proper size and type of aircraft for application of top-dressing materials at 180-270 kg/ha (160-240 1b/acre) to hill country in New Zealand showed that large cargo-type planes, such as the ex-military Bristol Freighter or the U.S. C-47 (DC-3), capable of carrying 4 000 kg (8 800 lb) or more per load, would be the most economical. The plan was to fly these from central airports to the dispersal area with large payloads that would afford the most economical cost per unit of top-dressing applied. But it was quickly found that there were very practical, if not wholly operational, reasons why this system would most likely not be used. First of all, it was thought that the Royal New Zealand Air Force might continue furnishing aircraft and pilots, but this did not prove to be the case, so private financing of aircraft and equipment had to be arranged. Although lending agencies indicated their willingness to provide financing to the new industry, they were more likely to start off with a modest loan for a small DH-82 Tiger Moth rather than a C-47 at ten times or more the cost. Furthermore, it seemed much more logical to haul the superphosphate in land vehicles to the hill country than to ferry it by aircraft. Also, by building local hilltop flight strips at relatively high elevations, the small aircraft could literally fly "downhill" in dispersing their loads. Other elements such as bad weather, winds, and fog could be quite local, and a plane based 16-80 km (10-50 mi) away from the dispersal area might either be unable to leave its base or find bad weather at the site.

The third factor which frequently determined the early choice of aircraft for particular countries or areas was availability. Referring again to New Zealand, the Tiger Moth was available in sufficient numbers and was, besides, a very practical, well-built, and easily adapted aircraft; consequently, it became the logical choice until better aircraft were made available in sufficient numbers to take over the job. Its replacement, the Fletcher FU-24, originally imported from the U.S.A. but later built in New Zealand, originally carried about twice the load of the Tiger Moth, and as larger engines were installed, this capacity was nearly doubled again to about 1 088 kg (2 400 lb). Lastly, this small, highly manoeuvrable aircraft was much easier to fly at lower levels over the steep hilly country of the New Zealand rangeland.

For agricultural spraying, where usual dispersion is 30-50 l/ha (3-5 gals/acre) or kg/ha, the area covered is 8 to 10 times greater than for top-dressing, and there is much less need for large aircraft. Thus, because of greater manoeuvrability, lower total costs, and higher economic productivity, the most satisfactory choice of agricultural aircraft is in the 500-1 000 kg (1 100-2 200) lb load class.

Of course, there are obvious exceptions. The large air tankers such as the Canadair CL215 PBY and TBM are highly specialized for fire fighting, where heavier aircraft flying at greater altitudes can be used. At the other extreme, aircraft applying sprays for locust and vector control (ULV, aerosol types) at 1 l/ha (0.1 gal/acre) to as low as 50 ml/ha (3/4 oz/acre) are limited in their single flight duration by fuel tank size, not by chemical load. A Canadian survey of operational costs also indicated that for forest, vector, and nuisance insect (black fly) control, the smaller, older, rebuilt aircraft were more economical per flight hour than any of the new and larger aircraft. Since engine maintenance is a basic variable cost (see Chapter 11), and since the original cost (capitalization) as reflected in the depreciation allowance is the principal fixed cost, it is reasonable to expect that the cheapest operation would be one with a low investment in aircraft and engines requiring low maintenance costs. Thus radial engines, which cost considerably more initially and 25-30% more to maintain and overhaul, are not widely used on newer low-maintenance aircraft.

Therefore, the choice of aircraft today is basically governed by the same parameters as would have to be evaluated in the purchase of a farm truck. The hours of use per year must be high, and specialty equipment that limits multiple use of the vehicle must be avoided so that the cheapest cost per hour can be attained. As the investment in equipment is increased, the productivity must likewise go up in order to obtain the most efficient economic use.

Helicopters generally fall into the high initial cost and high maintenance category; hence they have not been able to compete in the past with fixed-wing aircraft on the majority of aircraft jobs. Nevertheless, high productivity can be achieved with the helicopter, particularly with reduced volumes and wide swaths and for such specialty jobs as orchard and vine spraying, where the downwash at low speeds can make helicopter use advantageous (Ogawa, Yates, and Kilgore, 1964). There remains little question that for most application work the fixed-wing aircraft has successfully met the competition of ground equipment and has been able to do a comparable job at less cost per unit of area covered. Ground equipment is still most widely used, however, on many jobs, particularly for orchards and vines and for precise placement of materials, such as spraying of roadsides and ditch banks, as well as under sensitive crops (e.g. grapevines) for weed control. In these areas of work it seems likely that helicopter competition for ground applicators could be keen and provide an impetus for increased use of helicopters, as in France on grapevines.

7. AERIAL EQUIPMENT FOR DISPERSING DRY AND LIQUID MATERIALS

A great variety of materials are being dispersed by agricultural aircraft on many crops, rangelands, forests, and other areas, including various forms of dry materials, such as dusts and granular baits and seeds, as well as a variety of liquid foams, suspensions, encapsulations, emulsions, and various solutions atomized into a wide range of particle sizes, from coarse sprays of 4 000 to 5 000 μm (microns or micrometres) in diameter down to airborne particles, or aerosols, under 25 μm in diameter. Dry materials may range from large granulars of 1 000-2 000 μm major diameter down to 200-400 μm (Table 5). Dusts are of aerosol size, usual specifications calling for 75-95% by volume to be less than 20-30 μm (Table 6). Seeds may be several thousand microns in size, while dry baits may vary in granular size range and be larger.

Another important item in dispersal equipment selection is the total volume to be used per hectare or acre. As might be expected, the total volume affects the choice of application equipment — pumps, transfer pipes, and nozzle sizes — and, of course, interrelates with particle size in controlling the distribution of applied materials — the number of particles per square centimetre (or square foot) that actually collect on plant surfaces or ground when a given application is made.

Since agricultural aircraft are called upon to distribute both dry and liquid materials, it is frequently desirable to devise a system that can readily be changed from one to the other. The hopper or tank is built into the aircraft, and in some cases structural members will actually pass through the tank. Thus it is customary to make the tank both liquid and dust tight and arrange for the bottom to be removable, making possible the change from dry materials to liquid systems as may be needed. Figure 6 shows a liquid installation.

and Figure 7 a dry materials arrangement with a ram-air spreader on the bottom of the tank. As can be seen, the tanks are mounted forward of the pilot and located over the centre of lift of the airplane, so that changing weight does not affect the aircraft trim. Figure 8 shows a helicopter installation; because the centre of lift is directly under the rotor, the tanks are split, with one located on each side of the engine frame. Since the tanks cannot be removed for cleaning, they usually have a large top opening for entering the tank, as well as filling it with dry materials. Tanks may be made of stainless steel, but recently fibreglass construction has proved to be quite satisfactory for all materials, except for a very few chemicals which "attack" the plastic of the fibreglass tank.

LIQUID APPLICATION SYSTEMS

Figure 6 shows a typical boom-nozzle liquid dispersal system, which can provide a wide range of application volumes and drop sizes, depending upon the type of nozzle or other atomizer used. The tank is shown fitted into a space forward of the pilot and aft of the engine. The emergency dump gate (J) can be controlled by a lever located in the cockpit. The input from the tank to the pump is located in the bottom of the tank, and a by-pass flow from the control valve (C) directs the liquid back into the tank, keeping its contents thoroughly mixed by recirculation from the pump and hydraulic mixing of the tank contents.

In Figure 6 the pump (A) is driven by a propeller (B) in the aircraft slip stream. A brake with a cable and control (E) in the cockpit is used to stop the rotation of the pump. In normal use, however, the pump is allowed to operate continuously, and when not spraying, the liquid is recirculated into the tank for agitation. The control valve (C) is usually operated by a positive cable-control lever in the cockpit, but it may be operated electrically or hydraulically. The cutaway view in Figure 9 indicates the manner in which the control valve functions. This is essentially a three-way valve, which means it has three ports and three flow-control positions, although the third position may not always be used. In the "spray off" position, shown here, the valve directs the flow from the pump back into the tank through a venturi, or contracting, section, which causes reduced pressure (due to ac-

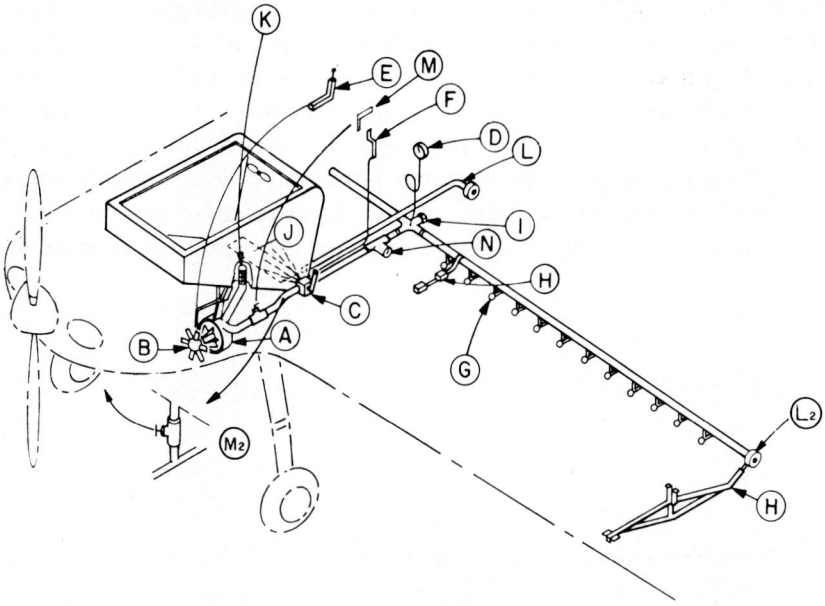

FIGURE 6. Cutaway diagram of a spraying system for a small fixed-wing aircraft.

celeration of the liquid) at the venturi throat and would empty the spray boom if check (one-way) valves were not placed at the nozzle. The flow-control screw shown at top right can be used to limit the flow back to the tank.

It should be noted that if several check valves at the nozzle were to fail, the flow would be greater than the suck-back device could handle; thus check valves should always be used and be kept functioning properly. Check-valve failure also permits the boom to empty, which delays the start of the spray. In the "spray on" position the crank is moved to the left, opening the line from the pump to the boom, at the same time closing off the flow to the tank. In practice the pump is allowed to run continuously while the tank has any liquid in it, in order to (a) maintain a vacuum in the boom to prevent dribble and loss of chemical while in flight or in turns and (b) provide recirculation agitation back into the tank. The third position of the valve, not always used, connects the tank to the boom (with lever turned to far right), in which case the tank may be emptied or filled through the end of the boom (L_2 in Fig. 6). Other types of control valves may be used, but the obvious advantages of the suckback system have made this a universal installation on most all aircraft sprayers. Valves are made with stainless steel or aluminium alloy bodies and plastic seats of polypropylene or Teflon, to reduce wear and ensure a close fit at all times. Failure of valves (and pump seats) usually occurs with wettable powders, which cause abrasion and wear of closely machined parts. In rare cases the chemicals attack the plastic parts and cause damage. However, daily cleaning

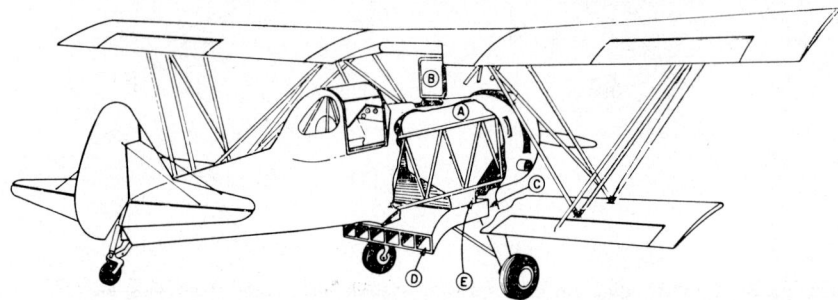

FIGURE 7. Cutaway diagram of a dry materials system with a ram-air spreader on a large (600 hp) biplane. Hopper capacity 907 kg (2 000 1b).

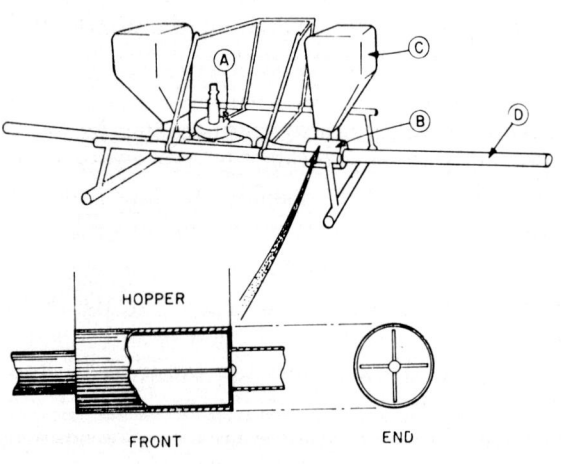

FIGURE 8. Air-driven dry materials spreader for a helicopter using electrically driven rotary feeder valves.

EQUIPMENT FOR DISPERSING DRY AND LIQUID MATERIALS 41

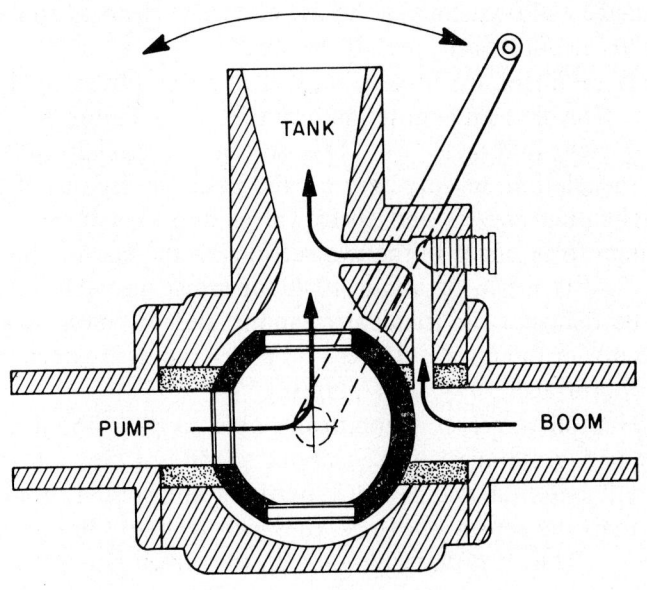

**3-WAY BALL VALVE
VACUUM POSITION**

FIGURE 9. Main control valve for aircraft sprayer, showing boom-vacuum position for positive spray shutoff. Check valves must be used at each nozzle with this three-way valve.

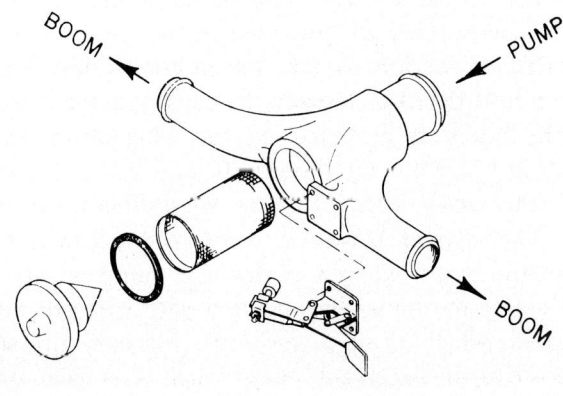

FIGURE 10. Liquid screen-filter used between pump and boom.

maintenance and occasionally the use of special cleaners and solvents, such as an alcohol wash, may be required.

Screens or filters are usually located at three places in the liquid systems. The first and most important is L in Figure 6, shown in exploded view in Figure 10. This screen is especially adapted to aircraft installation, serving both for the basic sieving out of particles that might plug nozzles and check valves and as a junction for the flow system from pump to the two sections of the boom. The screen (U.S.A. 25-100 mesh, or about 10-40 openings per centimetre) can readily be removed for frequent cleaning, and the mesh size can be changed to suit the nozzle orifices — smaller nozzles requiring smaller screens for protection against plugging.

The 25-mesh screen is theoretically able to pass about a 500 μm particle, and a 100-mesh screen about a 150 μm size. In practice, however, it is found that if many particles are present, even though smaller than the screen openings, they will tend to "bridge" or plug the screen. Thus it is desirable to keep the screen size just a fraction smaller than the nozzle orifice size. Smaller screens are frequently located at each nozzle (G in Fig. 6) just ahead of the orifice (Figs. 24a-c) and are the final point of clearance before the liquid is discharged past the nozzle orifice. In practice there is still a third point of screening, in the bottom of the tank at the point of entry into the pump. Here a large-opening screen of 2-3 mm (6-8 mesh) is used to keep large pieces of rock, iron, or other material from entering the pump. A practical system would then use a main screen of 25-50 mesh size and a nozzle orifice screen just under the orifice size.

The boom may be of stainless steel, brass, or aluminium alloy and either round or tear-drop or aerodynamic in section. Screw taps for the nozzles are provided in the boom, with an enlarged boss or threaded section to give the nozzle mount extra strength. Spacing can be ordered as desired; the usual spacing is from 15-30 cm (6-12 in). The boom is mounted on the wing structure with special fixtures (H in Fig. 6) — the inboard being close to the wing and the outboard farther away, because of the wing dihedral on the low-wing aircraft.

The valve at M is used to control the flow rate and thus the pressure on the boom when a centrifugal pump is used. This valve may be a single mechanically operated gate valve or an electrically operated valve, which is more desirable for easy pressure setting. When a positive displacement pump, such as a gear or roller type, is used, a valve and by-pass return to the tank is needed to control the boom pressure, as shown in Figure 6 (M_2). The pressure gauge is located at D, and pressure is adjusted on the boom during flight by setting the valve (M or M_2) to the gauge pressure desired. A flow metre of the recording type (N) is frequently used by organizations engaged in vector control and other large-scale, low-volume applications, where it is important to know the rate and amount discharged during an operation, especially when several small areas are treated successively during the course of a single flight.

The fitting shown at L is a quick-load attachment, which is coupled to the tank and contains a check valve operated by the male fitting on the end of the loader hose (see Fig. 59). This type of bottom loading permits rapid transfer of the liquid from the mixing tank to the aircraft with minimum exposure of loaders and pilot. The end of the boom (L_2) also may be fitted to the loader valve, and the transfer is then accomplished by placing the control valve (C) in the proper load position with the loader connected to the end of the boom (L_2).

Another type of spray installation is the bottom-attached, or belly, tank shown in Figure 11. This tank and pump assembly is attached to the bottom of the aircraft by quick-release fasteners, which permit a rapid jettison on the entire tank assembly if the need should arise. More importantly, the unit can be mounted on aircraft that are not primarily used for spray work. The entire unit is outside the fuselage of the aircraft, minimizing cockpit contamination and pilot exposure to spray material. The pump and propeller drive are as indicated for the previous system. The boom is hung from brackets and struts at G on the high-wing aircraft shown; bottom loading is provided for at B. The liquid system would be the same as illustrated in Figure 6.

Large aircraft, frequently used in military spray work and emergency programmes, are usually fitted with internally mounted tank and pump systems which can easily be installed and removed as may be needed (Lofgren, Mount, and Ford, 1970). Figure 12 shows such an installation, but with external pumps operated by the propeller wind drive system, and a full-wing-length boom system, as would be used for large-area spraying. For vector, tsetse fly, locust, and grasshopper control work the boom may be much shorter, since volumes would be small and the higher flight levels of the large air-

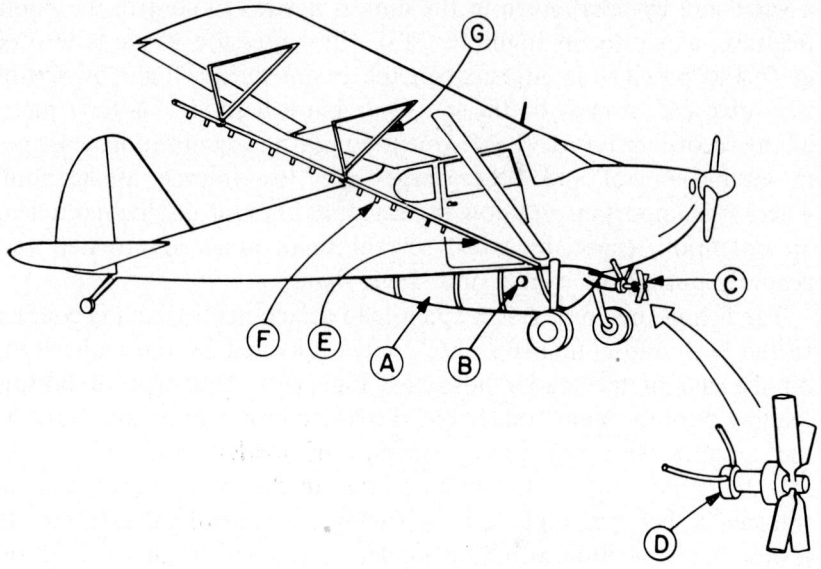

FIGURE 11. Spray system with quick-detachable belly tank on a Piper PA-18 fixed-wing aircraft.

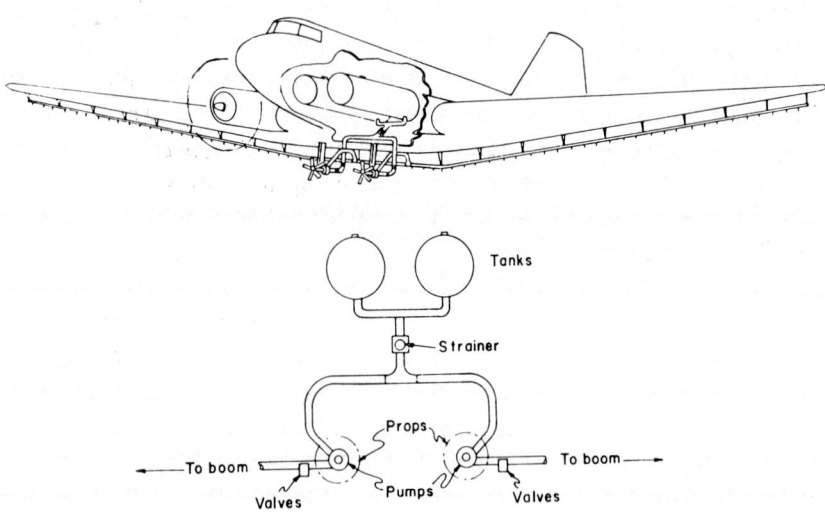

FIGURE 12. Double air-driven pump spray system with a capacity of 3 787 litres (1 000 gals) mounted on a C-54 (Dakota) airplane.

craft will cause the spray to disperse sufficiently for a wide swath pattern. Electric, hydraulic and even small gasoline engine drives may be used for the large aircraft sprayers, depending on convenience or availability.

Spray pumps

Spray pumps exist in a variety of types, but, as previously noted, the centrifugal type (Fig. 13) is most commonly used for application rates of 10-100 l/ha (approx. 1-10 gals/acre) and upward. The required capacity of the spray pump is then the total rate of nozzle discharge required at a given swath width and rate of travel. This relationship (see *Calibration of the aircraft applicator*) is as follows:

Q_T = Total output in litres per minute (gallons per minute)

$$Q_T = \frac{SQ_A V}{600} \text{ for litres per minute} \quad (a)$$

$$= \frac{SQ_A V}{495} \text{ for U.S. or U.K. gallons per minute} \quad (b)$$

The proper metric units (*a*) are:

S = Swath width in metres
Q_A = Applied rate in litres per hectare
V = Aircraft velocity in kilometres per hour

For the U.S. and U.K. units (*b*) use:

S = Swath width in feet
Q_A = Applied rate in gallons per acre
V = Aircraft velocity in miles per hour

Thus, for example, an aircraft flying at 150 km/hr (93 mi/hr) and applying 50 l/ha (4.34 gals/acre) on a 10 m (33 ft) swath would require a flow rate of 125 l/min (33 gals/min). Since the relationship is linear, one half the rate per hectare or one half the swath width would require one half the volume, also two times these rates would require twice the volume. This establishes some of the basic requirements for pumps; in practice, of course, and particularly in the case of positive displacement types (gear, etc.), a pump with larger than needed capacity will be chosen in order to absorb the loss in flow and pressure that occurs as the pump wears.

The power required to drive the pump is related to the flow rate, as explained above, and to the operating pressure of the boom. Normal operating pressure seldom exceeds 4.2 kg/cm² (60 lbf/in²), as higher pressures are less effective in attaining finer atomization (Fig. 28). The nozzles used (Fig. 24a) require a minimum pressure of 1.4 kg/cm² (20 lbf/in²) in order to spread out properly; thus the range of operating pressure is usually between 1.4 and 4.2 kg/cm².

Taking the rate of flow as calculated above, the horsepower required for the rate of water flow at a given operating pressure can be ascertained as follows:

$$hp = \frac{l/min \times kg/cm^2}{454 \times efficiency\ (decimal)} \quad \text{Metric}$$

or

$$= \frac{gals/min \times lbf/in^2}{1\,715 \times efficiency\ (decimal)} \quad \text{U.S.A.}$$

or

$$= \frac{gals/min \times lbf/in^2}{1\,429 \times efficiency\ (decimal)} \quad \text{U.K.}$$

For a flow rate of 125 l/min at 1.4 kg/cm² operating pressure the water horsepower (hp) is 0.385, and at 4.2 kg/cm² it is 1.16 hp; however, the power input required because of the low efficiency of the pump (25% for a small centrifugal pump and more than likely less for gear and other types) must be 0.385/0.25 (percent expressed as a decimal), or 1.5 hp, at 1.4 kg/cm² and 4.6 hp input at 4.2 kg/cm². If the flow rate per unit of area is doubled to 250 l/ha (10.7 gals/min), then at 4.2 kg/cm² (60 lbf/in²) the horsepower input to the pump would have to be:

$$\frac{250\ l/min \times 4.2\ kg/cm^2}{454 \times 0.25} = 9.3\ hp$$

Figure 13 shows a cutaway view of the commonly used centrifugal pump. These characteristically give high volume at low pressure, but with proper bearings and pressure-type seals they are very well suited to aircraft sprayer use. The diagram shows the air-driven propeller at the left (A), the brake system (B), the ball bearings (C), the pressure seal (E), and the liquid impeller (D). The latter is a closed-blade impeller; a variety of these are available for various purposes, but the closed blades have the best operational characteristics. The pump bodies are aluminium alloy with bronze or stainless steel impellers. Shown at the far right (G) is the input to the centre of the pump blade and discharge port at the top. The two different-size holes at the top and bottom indicate the so-called volute (expanding volume) principle, used to gain as much transfer efficiency as possible. The volume displaced by the centrifugal pump is related both to the velocity of the impeller and to the size of the impeller and the liquid-carrying channels. The pressure is related to the impeller diameter (peripheral velocity) and the shaft speed (rpm²). The practical limitation of around 4 000 rpm for an air-driven pump plus a pump size limitation means top pressure capability of around 4.2 kg/cm² (60 lbf/in²), as noted above. For higher pressures, higher speed or multiple stages (back-to-back) are used.

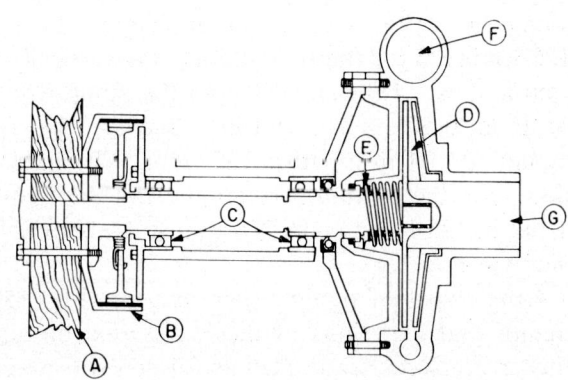

FIGURE 13. Cutaway diagram of an air-driven centrifugal pump, showing brake on the left and pressure type liquid seal on the right.

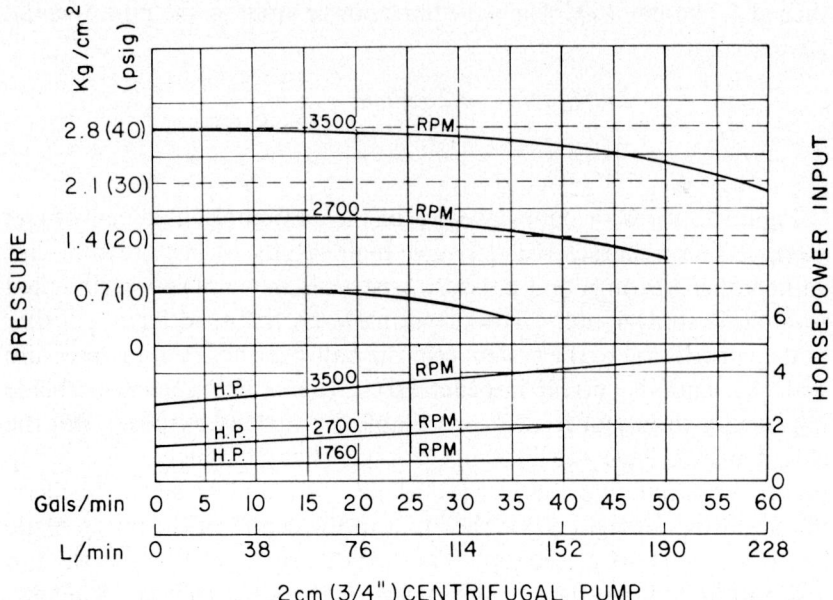

FIGURE 14. Operating characteristics (pressure versus flow rate) of a centrifugal pump with a discharge port 2 cm (3/4 in) in diameter. Shown for three different rotor speeds.

Figure 14 shows the operating characteristics of the centrifugal pump with a discharge port 2 cm (3/4 in) in diameter. The pressure is shown on the left-hand scale, the discharge along the bottom, and the horsepower required to drive the pump on the right-hand vertical scale. As can be seen, there are three rpm graphs each for capacity (top) and horsepower (below). The top graphs are read by moving along the bottom scale to the discharge value needed — for example, 125 l/min (33 gals/min) — and moving vertically to intersect one of the rpm graphs. Thus at 1 760 rpm the pump would produce a pressure of about 0.5 kg/cm^2 (7 lbf/in^2), and at 3 500 rpm it would give 2.6 kg/cm^2 (37 lbf/in^2), both at 125 l/min. To find horsepower at the same flow rate, the bottom rpm graphs and the right-hand scale are used. Thus, at 125 l/min and 1 760 rpm it is about 0.4 hp and at 3 500 rpm nearly 4 hp.

Other types of pumps, such as gear (Fig. 15) and roller (Fig. 16) pumps and variations of these, are used on aircraft either when (*a*) higher pressures are needed as for aerosol spraying or (*b*) the applied volume is greatly reduced, thus reducing the pump discharge requirement. Rotary (gear, roller, cam, etc.) pumps are characteristically low volume and capable of high pressure; thus small rotary pumps (Fig. 16) might be capable of 40 l/min (approx. 10 gals/min) at 21 kg/cm^2 (300 lbf/in^2). Rotary pumps are dependent upon metal-to-metal surface contact, require relief, or by-pass, valves (Fig. 6) in order to maintain constant pressure under varying loads, and are used to adjust the initial boom pressure. If a return-flow valve (see Fig. 9) is not used, the relief valve provides a by-pass for the liquid when the boom is shut off.

The rotary pumps have to be more precisely made, wear more rapidly, and in general do not fit aircraft requirements as well as the centrifugal type. It is customary to make rotary pumps of stainless or other highly corrosive-resistant steel. In some cases, as for the pump shown in Figure 16, the rollers are made of nylon plastic for long wear under abrasive duty.

Power sources for pumps

By far the most common power source for the spray pump is the direct propeller drive, with the pump and propeller placed in the slip stream of the propeller of the main aircraft engine, usually between the landing-gear wheels. The efficiency of such a system is very low. Not only is the pump itself rather inefficient at 25-35%, but the energy transfer from engine propeller to pump propeller is also likely to be on the order of 25-35% efficiency, which reduces the over-all efficiency to about 10-15%. Thus, if the liquid system requires 2.5 hp, an over-all efficiency of 10% means that 25 hp are actually required from the aircraft propulsion engine (Smith, 1969). Tests run on the horsepower developed by an air-stream propeller system installed on a Stearman 300 hp aircraft showed that, when cruising at 137 km/hr (85 mi/hr), the air speed in the propeller slip stream was 177 km/hr (100 mi/hr). With the usual 45 cm (18 in) six-bladed automotive fan, the power developed was around 4.5 hp. More elaborately designed wooden propellers achieved as high as 7.5 hp. For this reason, and also because of the need for better pressure control, other more expensive but better-engineered pump drives have been used. Figure 17 shows two hydraulic systems on an aircraft.

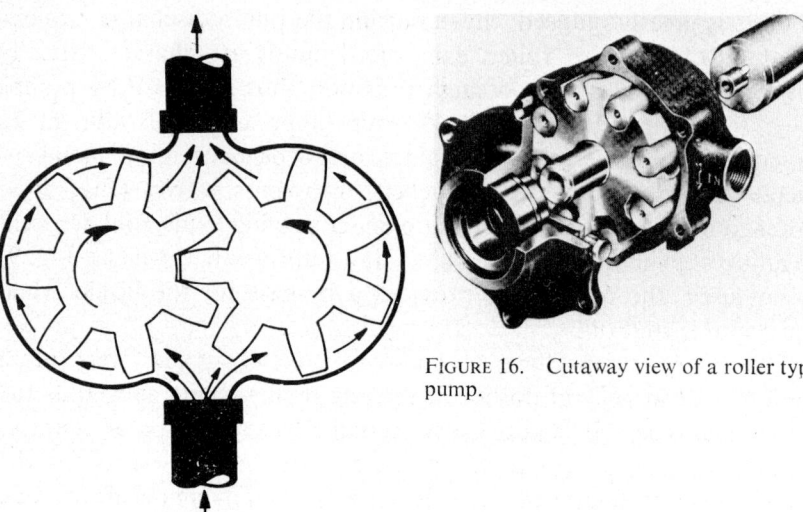

FIGURE 16. Cutaway view of a roller type pump.

FIGURE 15. Cutaway view of an external-drive gear pump.

FIGURE 17. Aircraft sprayer with centrifugal pumps driven by hydraulic motor. Hydraulic pumps (not shown) are powered directly from main engine.

The schematic drawing in Figure 18 shows how the system is organized. Basically, a hydraulic pump is installed at a suitable point of power take-off from the aircraft engine (lower right). Hydraulic fluid is drawn from the storage tank and pumped past the pressure gauge and surge tank to the by-pass pressure relief valve. The relief valve is in turn controlled by the control valve (upper left), which is adjusted in the aircraft cockpit for desired pressure. A by-pass around the control valve is used as an on-off valve, so that the system can be stopped and started again without readjusting the pressure. The relief valve feeds a fixed flow and pressure to the motor, which being a positive displacement type (Fig. 19) is speed-controlled by the volume of liquid fed to it. The excess (by-pass) oil from the relief valve, and thus the used oil from the motor, returns via the filter and cooler to the storage tank. Hydraulic systems of this sort operate at 105-210 kg/cm² (1 500-3 000 lbf/in²) oil pressure and are capable of 15-25 hp with relatively small packages (Fig. 19). The spray pump would

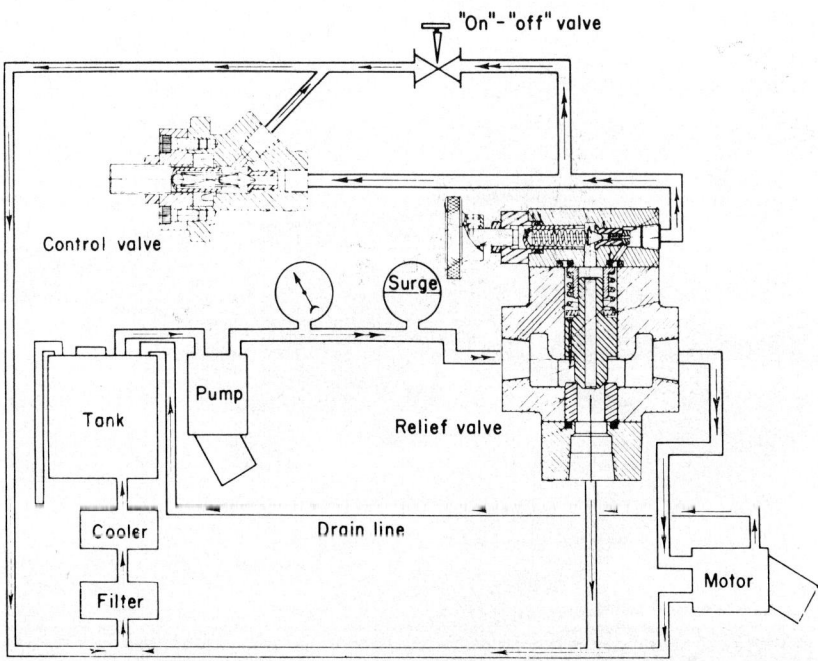

FIGURE 18. Schematic drawing of a hydraulic-powered aircraft spray system showing enlarged sections of control and pressure relief valves.

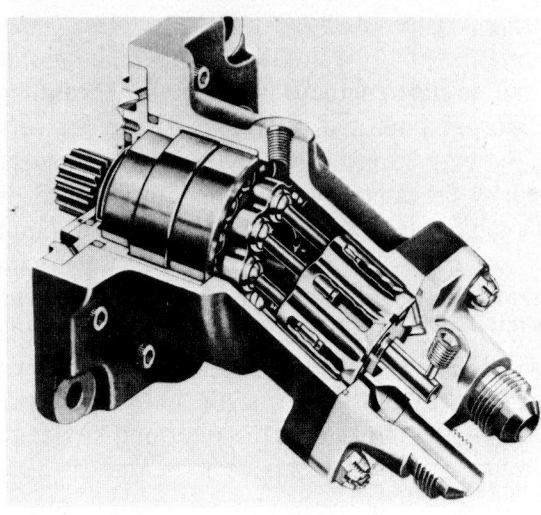

FIGURE 19. Cutaway view of hydraulic (oil-driven) piston type pump or motor for operating spray pumps.

FIGURE 20. Spray system with electrically driven gear pump for low-volume spraying mounted on a Piper Pawnee aircraft.

be either driven directly from the hydraulic motor or by a belt, chain, or gear system.

Hydraulic systems are very widely used on commercial aircraft, but drives of this sort have proved expensive to operate and not easy to design and install on old aircraft. The electric drives are probably the cleanest and easiest to use (Fig. 20). These have a good overall efficiency, but for requirements over 1-2 hp the usual 24-volt DC system becomes both expensive and physically large; for example, 5 hp or 3 730 (5 × 746) watts would require 155 amperes at 24 volts. Thus the electric drives are being used primarily for low and ultra-low volume applications where pump power is around 1 hp or less.

Helicopter sprayers are most often powered by another type of system: the power take-off drive (Figs. 21a, b). Here, the main propulsion engine must have a suitable PTO (power take-off) point, capable of the horsepower required (10-15 hp). The pump is then either directly connected through a clutch to this take-off point or may be belt-driven. Figure 21b shows a belt-driven pump on a fixed-wing aircraft. Belts provide the additional safety feature of protecting the gears in the engine in case the pump becomes locked. PTO drives are very economical and practical, and will probably be used more on all new aircraft in the future. As in the case of hydraulic and electric drives, PTO is not easily attached to old aircraft engines that were not designed for their use.

Spray atomizers

Many types of atomizers have been used on aircraft sprayers. The earliest were the rotary types, consisting of brushes (Fig. 22), propellers, disks and spinning wire cages (Fig. 23). More recent spinning types have used very fine sintered metal sleeves, small punched-hole screens, and a great variety of gauze and screening. The brush system (Fig. 22), developed in California in the 1930s, was most recently used in Canadian forest spray work. By far the most practical and useful of the spinners is the Micronair (Fig. 23), which has enjoyed widespread use, primarily in conjunction with low-volume spraying of large areas for locust and grasshoppers and with vector control of mosquitoes, black fly, tsetse fly, and the like. The Micronair AU3000, for example, is an excellent spinner type atomizer, which can be varied from 3 000 to 13 000 rpm (90 knots flying speed),

FIGURE 21a. Mechanical belt-driven centrifugal pump with electric solenoid clutch disengage mounted on a Bell 47AG helicopter.

FIGURE 21b. Mechanical belt-driven centrifugal pump on a Stearman fixed-wing aircraft.

FIGURE 22. Air-driven spray atomizer with spinning wire brush mounted on a fixed-wing aircraft.

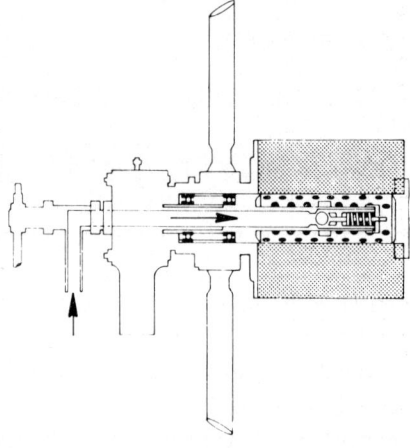

FIGURE 23. Cutaway diagram of a spinning screen (Micronair) type of air-driven atomizer.

depending on the pitch setting (15° to 50°) and type of impeller blade. The life of this unit is about 3 000 hours.

Various compressed gas systems have also been used to power liquid application systems. An air compressor, electric or engine powered, can be used in the aircraft, to furnish air power to force liquid out of an airtight system. In other instances, cylinders of compressed carbon dioxide or nitrogen have been used through suitable pressure reduction systems to apply pressure to gas-tight containers and force liquid to boom nozzle or spinner atomizer systems. Care must be taken to ensure that tanks are tested and approved (official inspection is usually required) to withstand the pressure to be used.

By far the most widely used and universally applicable atomizers are the hydraulic pressure types (see Fig. 24a). Here, the nozzle design is such that liquid under pressure fed to the nozzle and discharged into the air is broken up by the relative velocity of air and liquid. This is further affected by the aircraft slip stream velocity, as will be shown.

There are many good treatises on the atomization process and on the relationship of liquid characteristics, viscosity, density and surface tension to nozzle size, liquid pressure, and finally to the drop size produced. The Frazer equations (1958) are widely used, and Yeo (1952, 1961) has adapted these to the special conditions for aircraft spraying, where atomization is a function not only of the hydraulic pressure or centrifugal force (of a spinner), but also of the air shear of the high-speed slip stream and the discharge angle of the liquid to the air.

Figure 24a shows four hydraulic pressure nozzles and, at the right, a twin-fluid liquid and air nozzle for very fine atomization. The first nozzle on the left is the jet or solid stream type used for very coarse sprays, such as the phenoxy herbicides. The solid jet-spray pattern is produced by the spray discharge passing through a single orifice, as shown in cross section. Orifice size is designated by the Spraying Systems Co. in 64ths of an inch; thus a D2 and a D3 are orifices having diameters of 2/64 and 3/64 of an inch (0.8, 1.2 mm).

The second nozzle is the very widely used hollow cone type, identified by the spray pattern that it makes. The whirlplate in back of the orifice, shown in cross section, causes a spinning of the discharged liquid to produce a wide cone — the width being a function of the operating pressure and the size of the whirlplate in relation to the orifice.

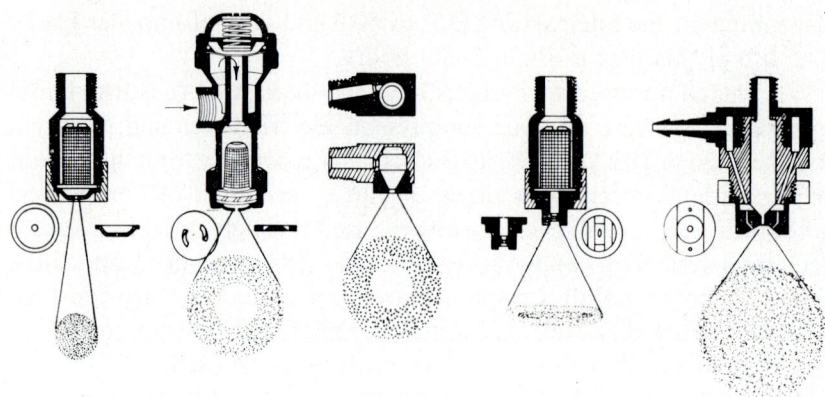

FIGURE 24a. Hydraulic-pressure nozzles. Left to right: (*a*) jet or solid stream for large-drop, low-drift spray; (*b*) the most commonly used hollow cone spray, showing whirlplate, which also uses a plate orifice similar to that of *a* (note also diaphragm type spring-loaded check valve); (*c*) side-entrance hollow cone nozzle; (*d*) fan type nozzle; (*e*) twin-fluid (air and liquid) nozzle used for producing aerosols.

Thus the smaller the whirlplate number — identified by the Spraying Systems Co., for example, as 13, 23, 25, 45, and 46 — the smaller the whirl openings and the greater the spin given to the spray. The smaller whirl by producing a wider cone angle also emits a finer spray. The diaphragm check valve at the top of the second nozzle, when used in conjunction with the boom suck-back system (Fig. 9), gives positive shut-off control of the spray equipment. The check valve consists of a spring-loaded chemical-resistant diaphragm. When the boom pressure exceeds the spring pressure of 0.2-0.56 kg/cm^2 (3-8 lbf/in^2), the diaphragm is lifted, allowing the spray to pass the diaphragm and go down past the screen and out the whirlplate and orifice.

The third nozzle from the left in Figure 24a is another form of centrifugal or cone nozzle, called a non-clog nozzle since its orifice is much longer than that of the hollow cone type with whirlplate. Spin to the discharged liquid is obtained by the tangential input to the nozzle. But it is not as versatile as the previous nozzle insofar as drop size is concerned, and hence is not so widely used. The fourth nozzle in the illustration is the flat fan type, widely used on ground sprayers as well as for reduced-volume applications with aircraft sprayers. Here, the orifice has a double milled slot, one on each side at 90 degrees to each other, causing the discharged liquid to fan out in a long, narrow pattern as shown. As in the case of the cone nozzles, the wider pattern also produces the finer spray. Spraying Systems Co.'s designation for fan nozzles indicates the first two or three numbers (such as 80 or 110) as the included angle of the fan when operated with water at 40 lbf/in^2. The next number is the whole-number flow rate of the water in gallons per minute at 40 lbf/in^2, and the last number is the decimal portion of flow rate. Thus 8004 is an 80-degree fan at 0.4 gals/min discharge, and 80005 is an 80-degree fan at 0.05 gals/min discharge, both for water at 40 lbf/in^2.

The last nozzle as mentioned earlier, is the twin-fluid, very fine atomization type, capable of true aerosol spraying (under 25 μm VMD). Here, the liquid passes through the centre of the nozzle, and the air enters in two jets which impinge on the liquid from each side. The effect is to produce a cone dispersion of very finely atomized spray, down to 15-20 μm VMD. However, the liquid orifice is so small that plugging of these fine nozzles becomes a serious problem. Nevertheless, these are about the only practical means known for producing sprays below 25 μm VMD from an aircraft sprayer.

The drop size produced by these various hydraulic nozzles is related to the nozzle type and design, primarily the fan or cone angle produced and the orifice size. Since the nozzles are used on aircraft in an airstream, Figures 24b and 24c illustrate how these are related. Figure 24b shows a hypodermic-needle atomizer (Microfoil, patented by Amchem Corporation) utilizing an air-foil section and single-drop type breakup as opposed to the filaments or fans of the four hydraulic nozzles in Figure 24a. In an airstream of less than approximately 95 km/hr (55/60 mi/hr) the drops emitted from this device with an internal needle diameter of 0.33 mm (0.013 in) are in the size range of 800-1 000 μm and practically constant in size. The device is used primarily for brush and weed control with the translocated phenoxy.

The last nozzle in Figure 24b is the jet type, which directed as shown, zero degrees to the airstream, produces drops of 600-800 μm VMD. Because of its atomization characteristics, this nozzle produces a wide range of drop sizes, or a log-normal distribution, characteristic of atomizers, with the exception of the Microfoil and certain other types operated under special conditions. The distribution is customarily plotted on log-probability paper (Fig. 50).

Figure 24c shows two other positions of the spray discharge — on the left, directed at zero degrees or with the airstream; on the right, at 90 degrees or down across the airstream. The angle of the nozzle may also be defined in terms of its direction in relation to flight direction. Nozzle direction can, of course, be at any angle in between, but the optimum effect will be seen to occur when directed at 90 degrees down or across the airstream. Figure 25 shows some generalized average drop sizes (see Chapter 10) in relation to hydraulic

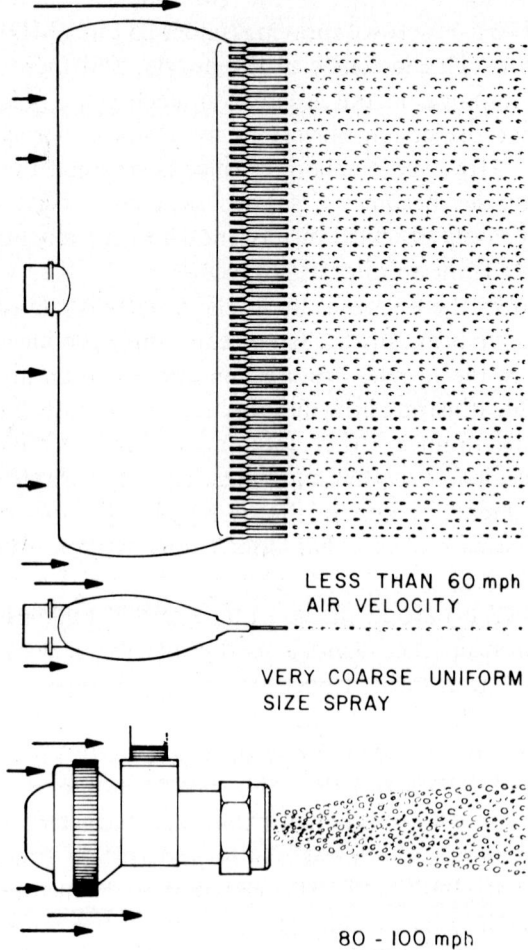

FIGURE 24b. Above: Low-turbulence microfoil type nozzle for uniform large-drop sprays (800-1 000 μm low drift). Below: Solid jet nozzle recommended for large-drop-size sprays (600-900 μm VMD), with diaphragm-type check valve.

EQUIPMENT FOR DISPERSING DRY AND LIQUID MATERIALS

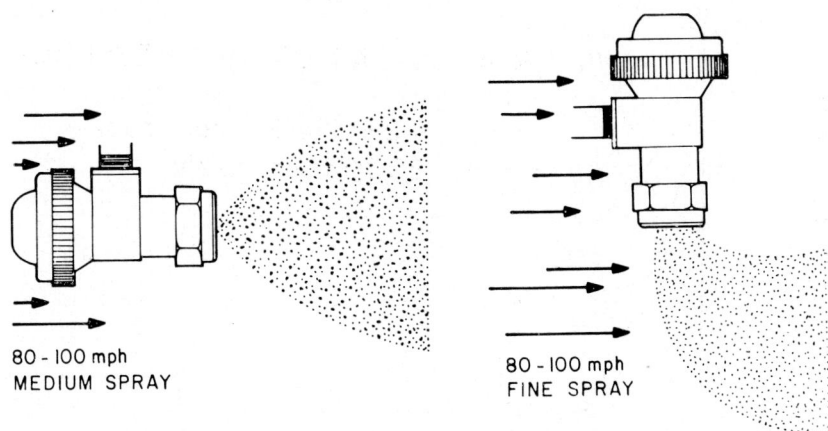

FIGURE 24c. Left: Hollow cone nozzle directed with airstream for a medium-drop-size spray (400-500 μm VMD). Right: Hollow cone or fan nozzle directed 90 degrees toward the airstream for fine to medium sprays (120-350 μm VMD).

pressure, with or without the 161 km/hr (100 mi/hr) airplane airstream, for two rather fine spray nozzles: a D2-13 hollow cone and an 80005 fan. The two top curves on the graph are for hydraulic operation only (Spraying Systems Co., 1968), and the two bottom curves show the effect of the 161 km/hr airstream on the average drop size produced. For example, at 7 kg/cm² (110 lbf/in²) the drop size with hydraulic pressure only from a D2-13 cone nozzle is seen to be about 150 μm VMD, while for the 80005 it would be about 215 μm. Increasing the pressure to 14 kg/cm² (200 lbf/in²) reduces the drop size to about 100 μm for the cone nozzle and to 170 μm for the fan; but placing these in a 161 km/hr airstream and directing their discharge down and forward into the airstream at 45 degrees gives a 90 μm VMD for the fan and 85 μm for the cone. Increasing the discharge pressure to 14 kg/cm² does not reduce this much further, since the energy requirement for further breaking up the drops becomes very high. It must also be noted that increasing the pressure of the liquid system for the cone nozzle operated without the airstream does not have much additional effect after pressure reaches 18-21 kg/cm².

Figure 25 shows the atomization of water which has a viscosity of 1 centipoise (cP). Although surface tension is high for water compared to petroleum oils (Table 8), viscosity appears to have a still

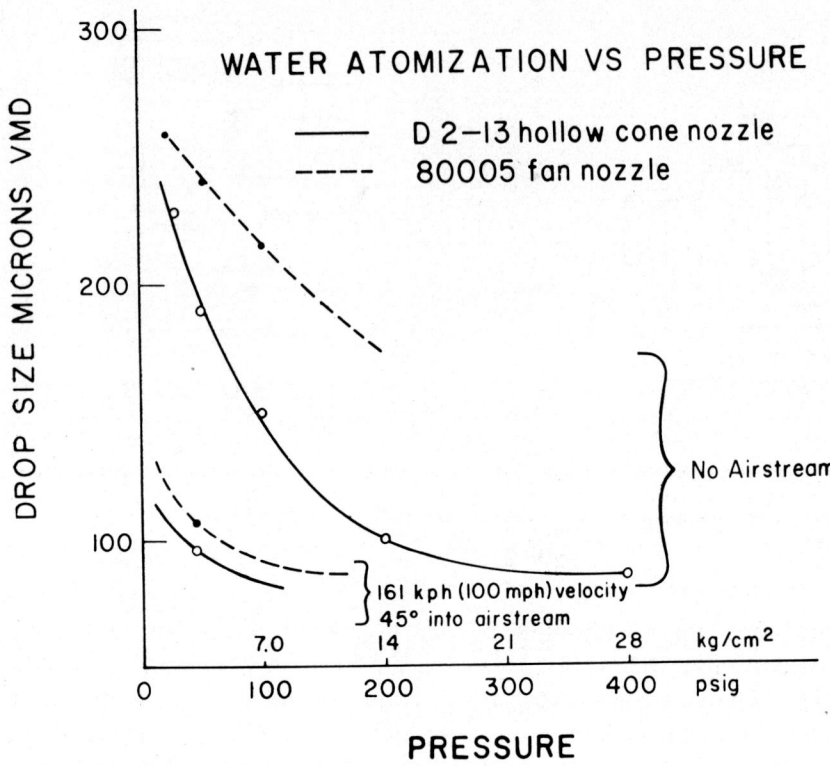

FIGURE 25. Drop size versus liquid pressure for a hollow cone and a fan nozzle. Upper set in no airstream, lower set in 161 km/hr (100 mi/hr) airstream directed at 135 degrees (against) airstream.

greater effect, and as viscosity is increased, average drop size also increases. This is illustrated for spinner nozzles in Figure 29, where the lower graph curve is for 10 cP diesel fuel, and the upper curve is for 45 cP malathion. A significant change in drop size can be seen even though the spinners were operating at about the same peripheral speed. Data from other observers (e.g., W. Maas) show that for an 8002 fan nozzle the change from diesel fuel at 10 cP to malathion at 45 cP causes a change in drop size from 137 μm to about 200 μm VMD.

Figure 26 shows the effect of the discharge angle to the airstream for two types of nozzles: a D6-46 hollow cone and an 8004 fan type. The D6-46 is operating with water pressure at 2.8 kg/cm^2 (40 lbf/in^2) and in a 161 km/hr (100 mi/hr) airstream. The 8004 fan is operating

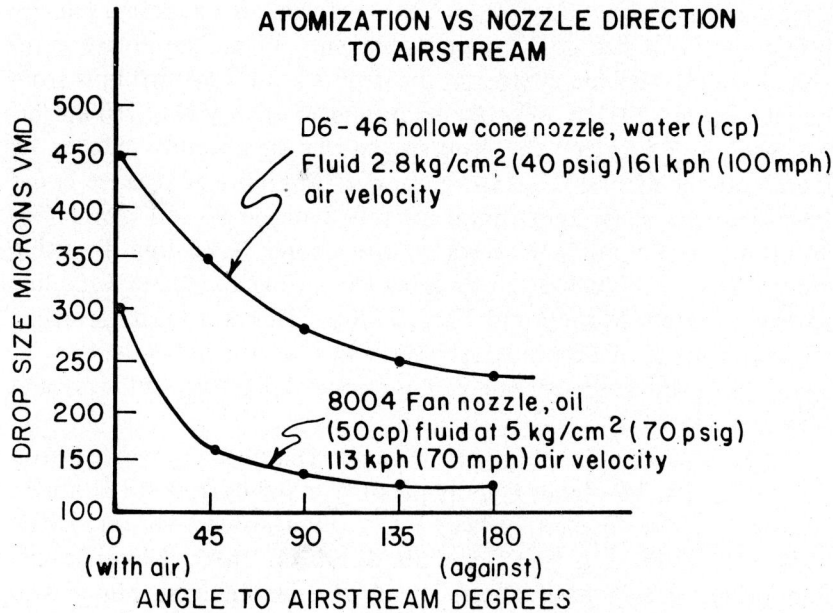

FIGURE 26. Atomization (drop size) versus spray discharge direction to the airstream for a hollow cone and a fan type nozzle.

with 50 cP fluid at 5 kg/cm² (70 lbf/in²) and in a 113 km/hr (70 mi/hr) airstream. As can be seen, the drop size for the D6-46 drops from 450 μm VMD when directed with the airstream to about 275 μm at 90 degrees and to 250 μm at 135 degrees (45 degrees into the airstream). The largest change can be seen from 0 to 90 degrees. Beyond 135 degrees little change takes place in the atomization. Although the fan has a higher-viscosity fluid, the much smaller orifice and the wider spread of the liquid fan cause a much smaller drop size to be produced by the fan nozzle.

Figure 27 indicates the effect on drop size when using an 8002 fan nozzle at two directions to the airstream — with and against, or 0 to 170 degrees — and when the air velocity is also changed from 0 to 300 km/hr (180 mi/hr). The fluid was 50 cP petroleum oil, and two liquid pressures are shown: 7 and 3.5 kg/cm² (100 and 50 lbf/in²). The effect of directing the discharge with the air is seen in the increase in drop size with velocity, until the airstream and liquid discharge rate are about equal (80 mi/hr at 50 lbf/in²). At this point one would expect drops of about 600-800 μm VMD to form, but increasing velocity brings the drop size down again as shown. When the nozzle is directed almost into the airstream, the drop size at 7 kg/cm^2 falls from about 175 μm VMD at 80 km/hr to around 75 μm VMD at 250 km/hr. Likewise, at 3.5 kg/cm^2 the drop size can be significantly reduced by increasing the airspeed. Thus, high-speed aircraft can atomize liquid finer than the usual agricultural aircraft flying at 90-150 km/hr (56-94 mi/hr), as the military vector-control people have done by using planes flying at 320-480 km/hr (200-300 mi/hr) to obtain very fine sprays (Lofgren, Mount, and Ford, 1970). This is also the principle of the venturi accelerator nozzles used in East Africa and in the U.S.A., again for producing extremely fine sprays (Parker, Collings, and Kahumbura, 1971).

Figure 28 shows the effect of the orifice chamber on average drop size. With a 45 whirlplate a spray pressure of 2.8 kg/cm^2 at 161 km/hr (100 mi/hr), the smallest orifice, a D1, gives around 90 μm VMD, while a D8 produces a drop size of about 190 μm VMD. Thus it is seen that orifice size in itself affects the drop size obtained, along with other factors of whirlplate size and nozzle design.

Table 9 shows some drop sizes obtained from twin-fluid nozzles. As can be seen, an increase in liquid pressure causes an increase in both flow rate and drop size. Also, increases in air pressure result in smaller drop sizes. The liquid flow must be very small in order to obtain fine atomization with these nozzles. At atmospheric pressure and standard temperature and humidity, air flow ranges from 0.02 to 0.08 m^3/min (0.93-2.8 ft^3/min) per nozzle.

Various spinner type atomizers have been used on aircraft, generally driven by small propellers as shown in Figures 22 and 23. Hydraulic and electric drives can also be used, but at higher cost per unit. The drop size can be decreased within about the same limits as with hydraulic nozzles: 75-100 μm VMD. Although very high peripheral speeds and very low flow rates might reduce this some, the practical limits of increased breakdown at high velocities restrict this as a further size-reduction technique. Also the flow limits are quite restrictive if small drops are to be formed. In Figure 29 the relationship of drop size to flow rate and peripheral (screen surface) velocity can be seen. Two actions affecting drop size take place as flow throughput is increased: (*a*) the spinner velocity drops, and (*b*) the screen surface must pass more fluid. Both of these actions make the average

FIGURE 27. Drop size versus air velocity for an 8002 fan nozzle in two positions relative to the airstream.

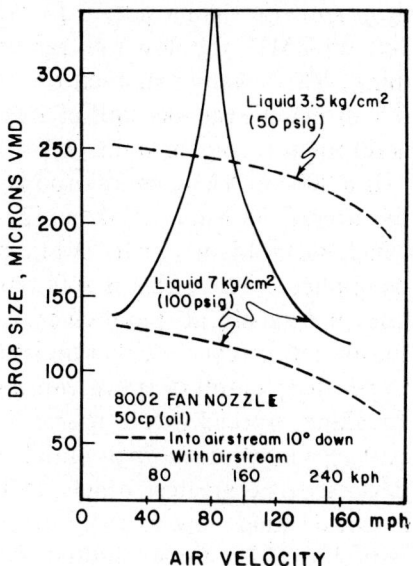

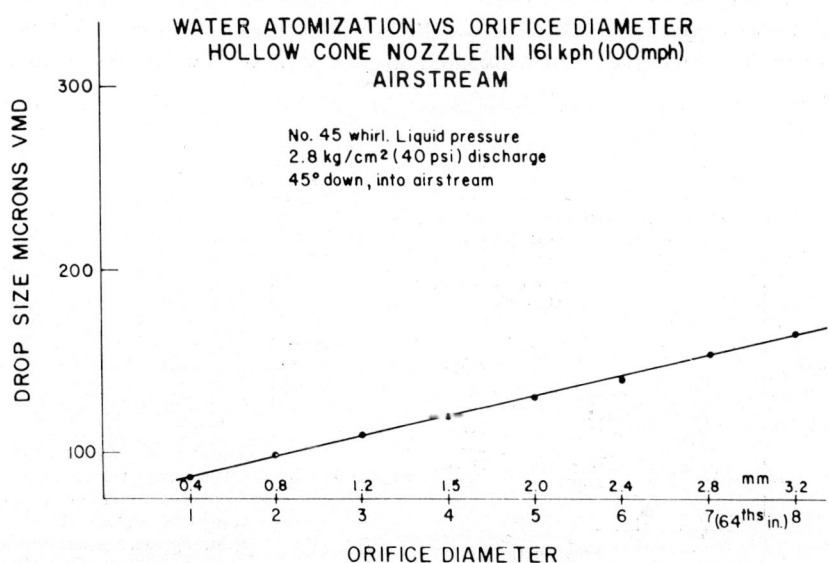

FIGURE 28. Drop size versus nozzle orifice diameter for a hollow cone nozzle directed 135 degrees toward the airstream.

drop size rise significantly. In order to keep atomization at around 100 μm VMD, the flow rate cannot exceed about 15 l/min (3.4 gals/min). This means that even on a 30 m (100 ft) swath at 136 km/hr (85 mi/hr), the rate per unit of area cannot exceed 3.8 l/ha (0.41 gals/acre) for two spinner units or 7.6 l/ha (0.82 gals/acre) for four.

If a 200 μm VMD spray drop is acceptable, then with four units on the aircraft, 38 l/min × 4 or 152 l/min (40 gals/min) can be obtained, which on a 15 m (50 ft) swath permits 178 l/ha (4.7 gals/acre) to be applied. While this is sufficient for many aircraft applications, it does not satisfy all requirements and would limit the usefulness of the aircraft sprayer. Other undesirable characteristics are (*a*) inability to get sharp cutoff of spray, causing dribbling in turns; (*b*) application streaking, especially as drop size is increased; and (*c*) wear on spinning parts. These, plus a high initial cost, limit the acceptance of spinner devices. The greatest benefit to be obtained is in the relatively unrestricted fluid flow as there are no small orifices to be plugged by particles in the spray solution. The greatest use of spinning devices is at low and ultra-low volume for large-scale vector, locust, and tsetse control programmes. A narrowing or reduction of the drop-size spectrum or range is claimed for the spinning device (Sayer, 1969), but this appears to be rather minor (although desirable) and related basically only to operation at a very low flow rate. In this case large-drop formation can be restricted at one end of the spectrum, and

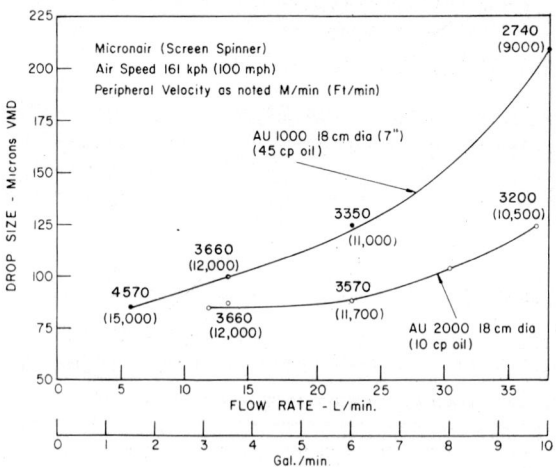

FIGURE 29. Atomization (drop size) versus liquid flow rate, and at peripheral velocities shown for the spinning wire-gauze (Micronair) devices. Data for two liquid viscosities also shown.

small drops may also be restricted due to the fracture limitations that become increasingly evident as drop size is reduced by any atomizer.

Aerodynamic drag in spinners — while not as serious as that caused by the spray pump located close to the airplane propeller — is nonetheless significant, and the loss at even a modest 10 hp per unit or 40 hp for four units renders it an inefficient system.

Several other types of atomizer have been used on aircraft, such as impact nozzles and the air-shear device. The first are not widely used because of the small orifice requirements and a tendency to drip after the spray is shut off; the second are largely experimental and have basically the same limitations as spinners because only low flow rates per nozzle can be used (Parker, Collings, and Kahumbura, 1971), and high aerodynamic drag of multiple units in the slip stream becomes restrictive.

Pipe and boom size

The size of fluid-transporting pipes and hoses in the aircraft should be adjusted to the expected flow rates. Thus for high-volume systems, up to 570 l/min (150 gals/min), the input to the pump should be 6.3 cm (2.5 in) in internal diameter and the discharge 5 cm (2 in). The boom for a boom-nozzle system should have an internal diameter (ID) of not less than 3.8-5.0 cm (1.5-2 in). For a lower flow rate the pipe and boom size can be reduced, but since the flow velocity is a function of area, and this in turn is related to the diameter squared, any systems with a flow rate above 133 l/min (35 gals/min) should not go under 2.5 cm (1 in) ID for pump discharge and transport lines as the pressure drop would become excessive.

When low and ultra-low volume applications are made, the pumps and flow lines are decreased accordingly. Thus at 19-38 l/min (5-10 gals/min) the pumps and lines can be decreased to 1.3-2.0 cm (1/2-3/4 in) ID. Since the 1.27 cm (1/2 in) line is a practical size to use, there is little need for using smaller pipes even for lower flow rates.

High viscosity sprays such as those produced by inverted emulsions (see Table 8), cellulose, and swellable polymers require larger pump and transport lines, since flow through a pipe is greatly affected by viscosity (Butler, Akesson, and Yates, 1969). Systems designed for handling highly viscous sprays at low flow or shear rates require that the pump be located very close to or against the tank bottom, with a

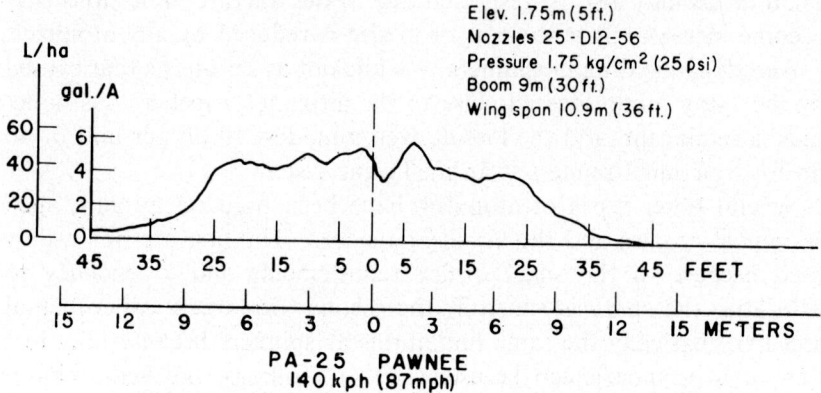

FIGURE 30. Single-swath spray distribution pattern for a Piper Pawnee aircraft.

6.4-7.6 cm (2.5-3 in) ID input line and a minimum 6.4 cm (2.5 in) ID discharge line and boom. This should be capable of handling viscous sprays up to 300-400 cP at flow rates normally used.

Pumps and pipe systems as well as controls, valves, booms, and nozzles for helicopters are much the same as those for fixed-wing aircraft. Figure 32 shows a helicopter sprayer, particularly emphasizing the large-diameter connecting pipe between the two side-mounted tanks. This pipe, 10-15 cm (4-6 in) in diameter, is used to

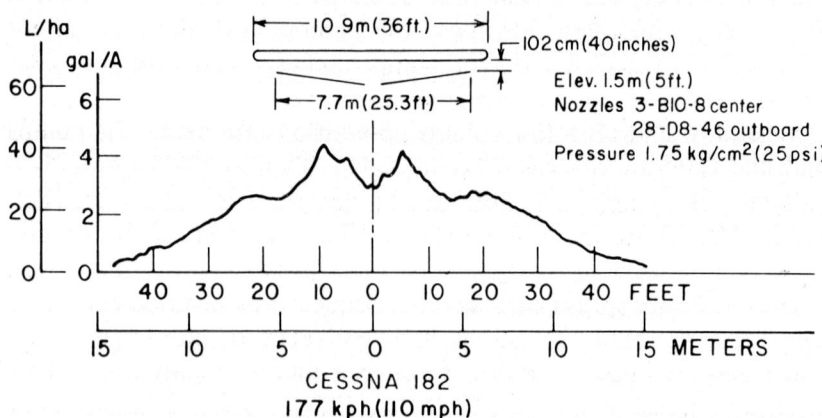

FIGURE 31. Single-swath spray distribution pattern for a Cessna 182 high-wing monoplane.

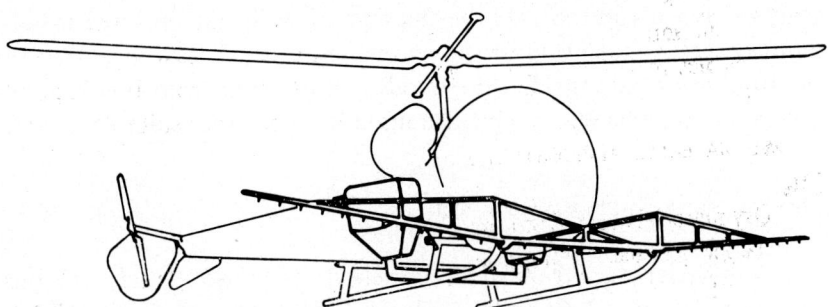

FIGURE 32. Spray boom on a Bell 47AG helicopter. Side tanks with 10 cm (4 in) diameter crossover pipe to maintain load level.

maintain liquid balance in the two tanks for lateral trim of the aircraft. The spray boom shown is about 9 m (30 ft), but it is commonly as long as 15 m (50 ft), therefore extending well beyond the rotor. But, since the helicopter can rise and land vertically, there is little chance of damage to the outer boom sections — which is not the case for the fixed-wing plane. Wider booms give wider swaths, but practical use limitations keep these at wing length on a fixed-wing aircraft and up to about 15 m (50 ft) in width on a helicopter.

Liquid-carrying lines, particularly those under pressure, should not be run into the cockpit or pilot area on any aircraft. On most air-

FIGURE 33. Ram-air type of dry materials spreader mounted on a Piper Pawnee aircraft.

craft these are kept outside the fuselage as well, but on large installations, such as the example shown in Figure 12, at least a portion of the lines must pass inside the aircraft. Protection from breakage of these lines must be ensured by armoured hose or other suitable means.

Dry material application systems

Dry materials vary from small particle dusts (75-95% under 20 μm) to larger granules, baits, and seeds of 1 000-5 000 μm (Tables 5, 6). Nevertheless, the application equipment has been very much the same for all forms of dry materials until quite recently, when spinner-type spreaders, developed initially for helicopter use (Baltin and Brandt, 1966), began to be transferred to fixed-wing aircraft and were found to be very effective in producing wide, low-profile swaths. These are relatively uniform for overlap matching and produce a good distribution pattern as the aircraft progresses across the field.

Dry dusts are seldom used due to high drift-loss potential, but the fixed-wing aircraft illustrated in Figures 7 and 33 and the helicopter in Figure 8 will handle dusts, as well as other dry forms. The fixed-wing aircraft utilizes the high air speed directly off the propeller slip stream, as much as 240 km/hr (150 mi/hr) on an aircraft moving at 161 km/hr (100 mi/hr). These are called ram-air spreaders and are widely used for moving dry materials into the wake of an aircraft in order to obtain distribution and spread.

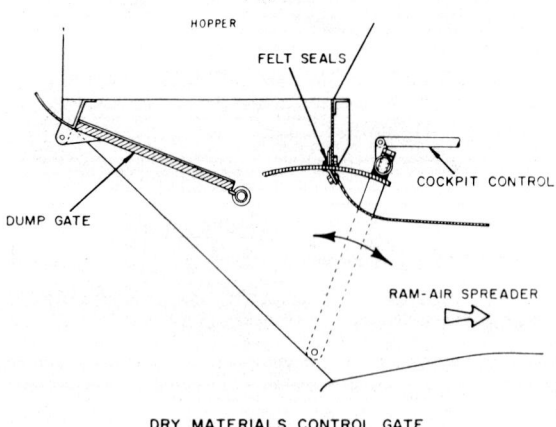

FIGURE 34. Cutaway of a sliding dry materials metering gate showing emergency dump gate at left.

EQUIPMENT FOR DISPERSING DRY AND LIQUID MATERIALS 69

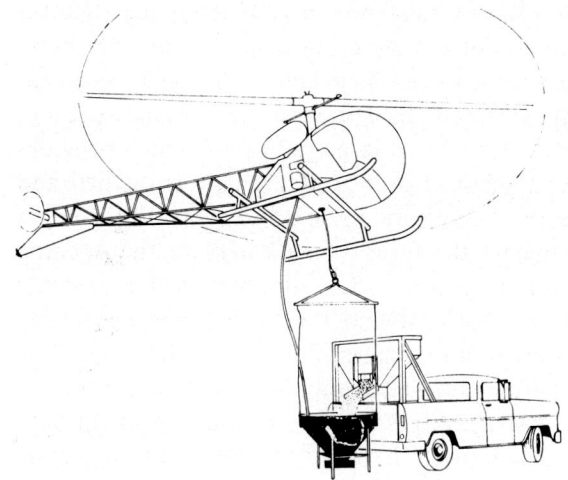

FIGURE 35. Quick-loading dry materials basket system with spinning disk spreader driven by hydraulic power from the helicopter. Loading may also be accomplished by detaching from empty basket and attaching to filled one.

FIGURE 36. Bottom view of an electrically driven spinning disk type of dry materials spreader on a Hughes helicopter.

The ram-air spreader may or may not have a revolving agitator located just above the throat of the metering gate. Figure 34 shows a typical sliding-gate metering device just below the tank, a quick-release emergency dump valve to the left, and the ram-air spreader below. The metering gate describes a large-radius arc which provides a certain amount of freedom or clearance as the gate is opened and closed. Seals of felt, leather, or plastic are used to close off the edges of the gate. The gate meters the flow by the extent of the opening and is also used to stop and start flow. Usually there is an adjustable stop located on the gate control in the cockpit, which permits setting the limit of the gate opening to a prescribed flow rate as determined by flight calibration and desired rate of coverage per unit of area.

The system in Figure 8 (page 40) is used in various forms on helicopters. In this instance a blower driven by the propulsion engine forces air out of two ducts past the bottoms of the side tanks on the helicopter and out the short booms. At each tank either a slide gate, such as that used on fixed-wing aircraft, or a revolving metering

FIGURE 37. Double-spinner dry material spreader, hydraulic-power driven, on a Stearman biplane.

gate is used. The positive feed afforded by the revolving gate is desirable, since lateral trim must be maintained on the helicopter and, hence, each tank must be emptied at the same rate. A small electric motor at each hopper drives the revolving gate. Changes in gate speed and an adjustable slide opening can both be used to control the rate of flow, all by remote electrical control from the cockpit.

Dissatisfaction with the ram-air spreaders led to experimental work in the U.S.A. and western Europe, as well as in the U.S.S.R., with the spinner type of dry-material spreaders (Azar'yan, 1966; Baltin and Brandt, 1966; Roberts and Smith, 1963). It was soon found that considerable power was required to move large volumes of material and spread them in the aircraft path. At 455-910 kg/min (1 000-2 000 lb/min), for a high rate of fertilizer application, the power requirement was 20-25 hp input to the spreader; for the helicopter flying at about two thirds the speed, the discharge rate and power could be reduced proportionally. The first units in the U.S.A. were somewhat as shown in Figure 35, but instead of the hydraulic-

FIGURE 38. Rapid bulk-handling system. Left to right: (*a*) Stearman airplane; (*b*) skip loader which holds one airplane load; (*c*) belt loader which transfers dry materials from bottom dump of (*d*) highway transport vehicle to skip loader.

FIGURE 39. Scoop type skip loader mounted on back of service vehicle. Dry materials are scooped up from open bulk storage by the loading bucket, and one aircraft load is transferred to the New Zealand Airtruk aircraft. *Photo Transvia, Ltd., Australia.*

power hose shown coming down from the helicopter to drive the spinning distributor, a 2-3 hp gasoline engine placed on the tank drove the spinner. The power in this case was insufficient to do a proper job at the flow rates required, so the hydraulic-drive system came into use for both fixed- and rotary-wing aircraft, as shown in Figures 18 and 19. If two motors are used, a flow-control device must be installed in the system in order to maintain uniform flow to each motor. Such a unit is shown on a fixed-wing aircraft in Figure 37. The two spinners revolve in opposite directions, throwing material outward from the front of the spinner in a remarkably uniform pattern. Deflector plates at the front of the spinners keep material from going forward and striking the landing gear and propeller, and also help guide the distribution. Flow control is obtained with the same type of gate used on the ram-air spreaders,

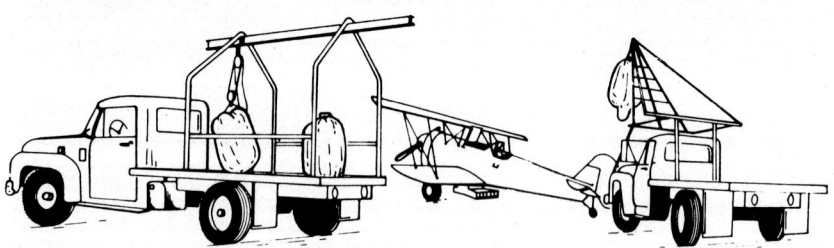

FIGURE 40. Unit-load bag system for dry materials. Left to right: (*a*) transport vehicle and loaded bags; (*b*) aircraft; and (*c*) bag handling vehicle.

although more elaborate auger systems have also been used and proved to be restrictive and power-consuming. Figure 36 shows a double-spinner PTO (power take-off) system driven from the main engine of a helicopter. Here, two spinners are used to enable distribution from the two tanks on each side of the helicopter.

Fertilizer and seed loading and spreading are massive handling operations, so bulk handling systems with large transport trucks, batch loaders, and labour-saving transfer systems are widely used. These also reduce loading time and permit rapid turnaround of aircraft used in such operations. The New Zealand skip loader and batch transfer tank placed on the front of a tractor or truck are commonly used there to transfer bulk superphosphate (Fig. 39). In California, systems involving the truck, loader, and transfer equipment shown in Figure 38 are commonly used; the large transport truck brings fertilizer from the factory direct to the field, and it is then loaded into the batch loader (one aircraft load) on a belted transfer

FIGURE 41. Dry material applicators. Left, above and below: View and cutaway of the Razak forced-air distribution system. Right, above and below: View and diagram of Mississippi project air-power spreader system. *Source: Mississippi State University.*

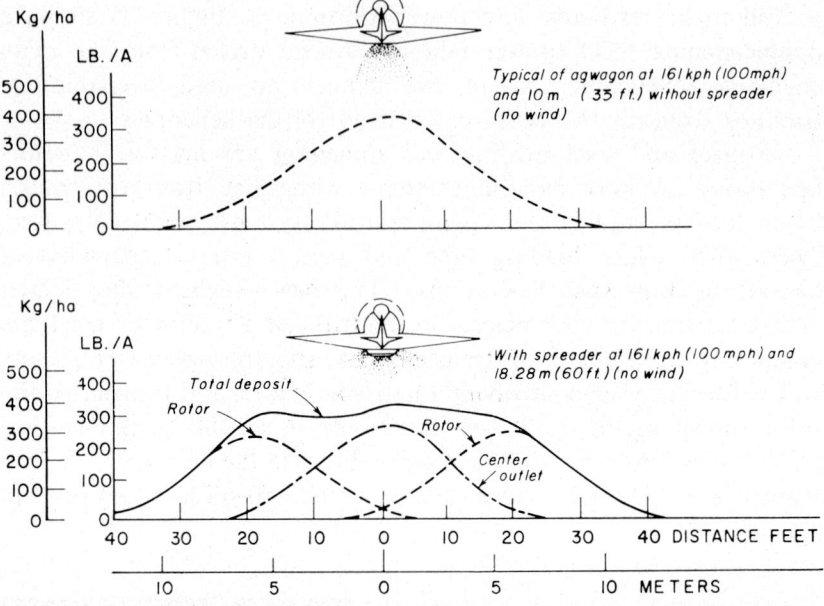

FIGURE 42. Top graph: Cessna AgWagon spreading dry granular materials without spreader. Lower graph: The same with addition of the New Zealand type rotary cylinder spreaders on each side.

elevator and goes from the batch loader to the aircraft. It is highly desirable to have a fixed quantity in each load transferred to the aircraft, as an accidental overload could cause an unexpected long take-off run and a possible crash.

Figure 40 shows still another system used for facilitating bulk handling of materials for aircraft distribution. Here, the large bags, up to a single aircraft load each, are preloaded in a central area, placed on the truck, and hauled to the landing strip, where the boom truck picks up each bag and empties its contents in the plane.

Spinners and ram-air spreaders are the two basic systems used for the dispersion of dry materials. As noted, the principal shortcomings of the ram-air devices are the low spreading power available and high drag. Drag is greatly reduced with the hydraulic-powered spinners, but by far the greatest gain is in the power available to perform the spreading job. Other schemes for increasing available spreading power are illustrated by three other types of dry applicators shown in Figures 41 and 42. Figure 41 shows the Mississippi project air-powered spreader on the right (Roberts and Smith, 1963). Here, a blower powered by PTO from the main engine is used to drive air, in addition to the normal ram-air, into a spreader. The passages of the spreader are laid out to give uniform distribution of the discharged material. Another powered spreader shown in Figure 41 (left) uses a 125 hp auxiliary engine to drive a blower, which forces air through a duct, past the hopper discharge and then to ducts located in each wing. While each of these systems gives highly improved swath patterns and can handle large discharge rates very well, the added cost does not at present justify their use. The aircraft on the left is especially designed for a large load capacity, and the discharge of air over the wing tends to give additional lift or "boundary layer" effects. The last spreader shown in Figure 42 illustrates a new rotary-cylinder spreader design developed by engineers at Auckland University in New Zealand (Lee and Stephenson, 1969). The principle of operation is to spread the material in three patterns, as shown in the composite graphs at the bottom of the figure. Here, a portion of the material falls out of the centre section and is distributed in a central pattern between 5 m and 10 m (16 to 33 ft) wide. The two revolving-cylinder side spreaders move material out another 10 m, giving an overall usable or flagged swath of about 18.3 m (60 ft).

8. APPLICATION TECHNIQUES: PHYSICS AND TECHNOLOGY OF PARTICLE BEHAVIOUR

The basic application objective, either by ground rig or aircraft, is to put the material where it is most effective, whether on plant surfaces, directly on insects flying or resting on the ground, or into water. While it would be highly desirable to have all the material land on the target, with none escaping or missing it, this is not possible at present, especially with all the formulations and applications that aircraft are being called on to use and make. This is also the core of the problem of chemical losses during and following application. The problem of getting materials to the ground or into water is readily solved by the use of granular materials, but, as indicated earlier, there are not very many chemicals that can be used successfully in granular form. These are limited to only a few insecticides which can be placed in the ground and will translocate through plant systems to the insect feeding or sucking areas. A greater variety of herbicides are available for this use, primarily those applied before plant or crop emergence. After the crop is up, granular applications are not very successful or else will damage the crop. Control of certain insects, such as mosquitoes and gnats, can be accomplished very nicely by applying granular forms of chemicals to the standing-water breeding areas. Most granular materials are made by absorbing liquified pesticide chemical in the particles of granulars. By using various types of granular materials and pesticide formulations the rate of release of the pesticide can be adjusted to give longer- or shorter-term protection in soil or water. Another application is the control of the corn (maize) borer by applying granular insecticides which catch in the plant whorls and stay there long enough to give adequate control.

Granular size distribution

In Table 6 the size distribution of granular particles is shown according to U.S.A. mesh (number of openings per linear inch), the actual opening size or particle size in microns, and the average number of particles per unit volume. The last two columns indicate the average number of particles per unit area at 11.2 kg/ha (10 lb/acre). Even the smallest granules shown — the 30/60 mesh, or 520/246μm — have a high settling velocity of about 1-2 m/sec (4-8 ft/sec); hence the loss of these from the treatment area would be very low.

The particle distribution and density for three types of commonly used dusts and wettable powders are shown in Table 5. The three types of readily available talcs and clays (Watkins and Norton, 1955) are widely used for their high absorptive abilities. The particle or solid density does not vary greatly, being two to three times heavier than water. But the bulk packed density (or bagged density) varies considerably, as the different size range and shapes of materials pack differently. The range varies somewhat, but 50% of the particles will usually be under 10 μm in diameter and 90% less than 20-30 μm. Since particles up to 20 or 30 μm can be totally airborne, or carried indefinitely in the atmosphere, it is easy to see why dusts have a high air-transport loss from the treatment area when applied by either aircraft or ground equipment. In practice, however, dust particles tend to stick to one another and thus do not behave as totally individual particles.

The small particle size of dust gives a tremendous number of particles per unit of volume and hence very good crop coverage. The same would apply to small spray particles in the aerosol size range. But the deposit of the small dust particles is very poor, even less than for comparable liquid particles. This is why dust use is being reduced in most areas of the world today.

Application distribution of granulars

Typical application characteristics of granular materials are shown for fixed-wing aircraft in Figure 43. Here, ammonium sulfate materials of a coarse granular form were spread at different discharge rates from a Cessna AgWagon, using a ram-air spreader. The lower rates of 225 kg/ha (200 lb/acre) can be applied on a 10 m (33 ft)

APPLICATION TECHNIQUES: PARTICLE BEHAVIOUR

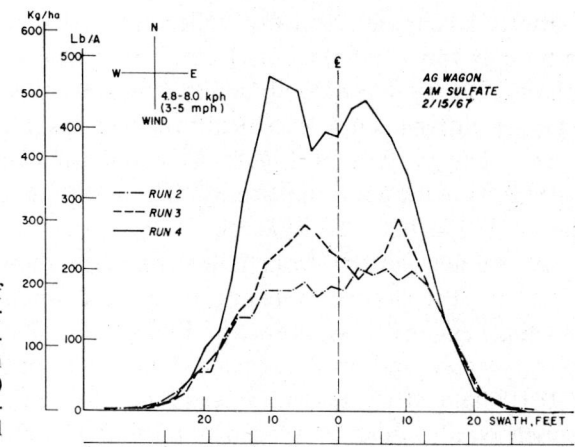

FIGURE 43. Graph of Cessna AgWagon ram-air type of dry granular materials spreader for two swath directions (lower curves) and at two rates of application. Wind velocity and direction are also shown.

swath path (flagged swath width), and a good matching of pattern will result. But, if the swath width is varied significantly, the peaks and valleys of the application pattern become very serious and cause bad streaking of the field. As the discharge rate is pushed upward to 450 kg/ha (400 lb/acre) the pattern narrows somewhat and the swath width should be reduced to 6 m (20 ft). Figure 44 shows a distribution pattern from a spreader of the spinner type. The total

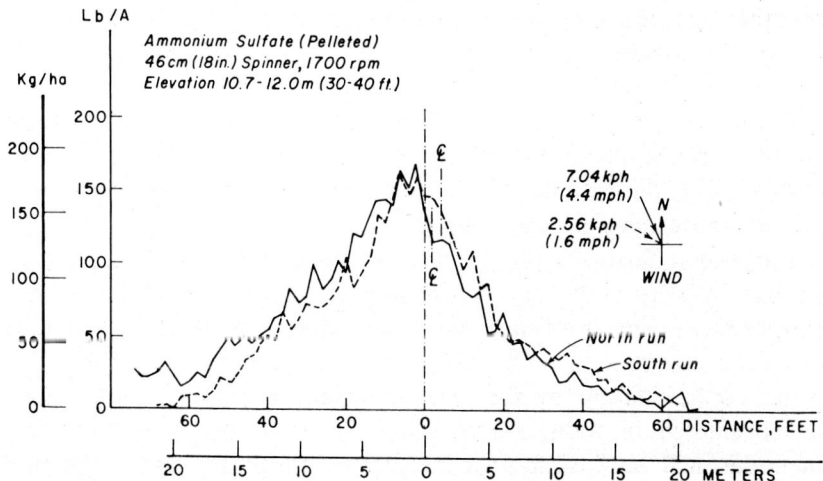

FIGURE 44. Graph of Stearman aircraft with single spinning disk type of dry granular materials spreader, showing two application directions.

pattern is characteristically wider than that of a ram-air device, although this advantage is lost if discharge rates are raised much above 225 kg/ha (200 lb/acre); however, the flatter profile of the spinner pattern is not as critical to swath spacing as with the ram-air type. The pattern of Figure 44 could thus be spaced at 12-15 m (40-50 ft), with good uniformity of coverage at an application rate of about 170 kg/ha (150 lb/acre).

A spinner pattern from a helicopter is shown in Figure 45. In this case the double spinners are spaced about 2 m (6.5 ft) apart on the helicopter, as shown in Figure 36. The distribution pattern is very good, and an 84 kg/ha (75 lb/acre) application rate at a 6 m (20 ft) flagged swath width could be obtained. At lower rates the swath width could be spaced out to 12 m (40 ft) or more. Again, the triangular pattern shown is to be favoured over the sharper-edge trapezoidal patterns, since it is common practice to adjust rate of application by altering swath width.

The top graph in Figure 42 shows the typical discharge pattern obtained when no spreader is used and the material is simply dropped out of a metering gate into the slip stream (Lee and Stephenson, 1969). Small aerodynamic (plough-shaped) devices are used on these aircraft, but no attempt is made to obtain wide swaths or uniform land coverage in the application of superphosphate top-dressing to New Zealand hill pastures. Even without any elaborate ram-air or other spreading device, it is seen that a rate of 225 kg/ha (200 lb/acre) is possible on a 6-10 m (20-33 ft) swath. This is a quite usual rate of application, and since the total area is being covered with the top-dressing, there is little incentive to try to obtain uniform swaths. When nitrogen is applied, particularly in high concentrations (40-45% N), the activity of the nitrogen and usefulness to the crop require more careful spreading.

Large-area fertilizer programmes to be carried out in a manner similar to that used for top-dressing in New Zealand have been proposed in the southeast Asian countries. While the technique could be of great value to crop production, it poses problems of water contamination by the materials being applied — which must be considered in today's agriculture. It is more likely that most materials will be applied in concentrated forms to high-production crops (rather than grassland and forests) and directed specifically to the crop area, rather than general broadcast coverage.

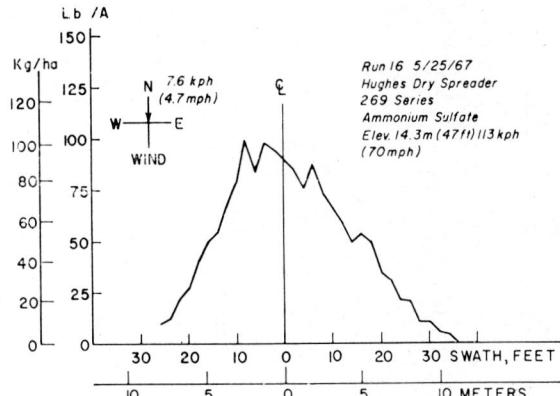

FIGURE 45. Graph of double spinning disk type of dry materials spreader on a Hughes helicopter (see Fig. 36).

Bait and seed spreading patterns

The distribution of baits on cereal bran, sawdust, and other carriers for grasshopper, rodent, bird, and other control is accomplished with the same type of equipment that is used for other dry-materials application. In the U.S.A. large aircraft (B-17 and DC3 or C47) have been used for this purpose (Messenger, 1953); but the use of large-volume baits has been largely displaced by low-volume sprays and wet baits with materials such as molasses or syrup as carriers.

For rodent and rabbit control by baiting, poisoned grain or other attractive materials can be applied by aircraft (Howard *et al.*, 1956). However, more recent work on rodent and bird control in Europe and Africa indicates that spray formulations are more useful for this purpose — rodents eating the spray-poisoned vegetation in one case, and birds being directly sprayed in flight in the other (Bauer, 1966; Lange, 1960).

Aerial seeding is widely used in agriculture for small grains and for wet seeding in rice paddys. The systems are similar to those used for fertilizer and granular chemical applicators, and the larger, heavier seeds will give a slightly wider distribution pattern, as can be seen in Figure 46. Here, the Cessna AgWagon with ram air spreader can be seen to do a very commendable job in applying rice seed on about a 12 m (40 ft) swath at 112 kg/ha (100 lb/acre). For very low seeding rates of small seeds, such as grasses and lucerne, a screen or slotted mask is placed over the meter-gate opening in the hopper bottom to reduce flow rate of the small seeds.

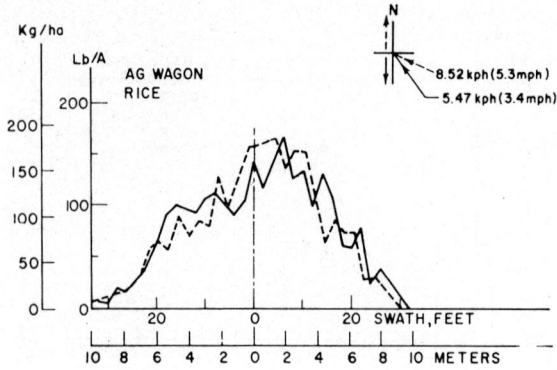

FIGURE 46. Graph of rice sowing with a Cessna Ag-Wagon.

Spinner-type spreaders will do equally well on seed spreading, although care must be taken to protect large seeds, such as rice, from the impact damage of a high-speed rotating blade. Whereas 46 cm (18 in) blades are operated at 1 600-1 800 rpm for granular materials, reduced speeds of 1 000-1 200 rpm may be required for seeds. Spinner blades are sometimes faced with rubber to reduce impact damage.

Figure 47 shows the cumulative effect of seeding rice with a spinner spreader. The individual pattern (single pass) is shown at the bottom of the figure. With a slightly asymmetrical pattern of aircraft discharge, the technique of to and fro applications (Fig. 48) in a 6.5 m (21 ft) swath shows considerable variation in the seed rate. If, however, the round-robin technique is used (Fig. 48), the overlapping of swaths with the same symmetry in pattern reduces the variation as shown in the second graph from the top in Figure 47. Theoretically, the pattern could have been further improved by flying on 5.8 m (19 ft) swath widths, as shown in the top graph.

The use of granular materials provides one of the safest means for aerial application, particularly of toxic chemicals. In the near future a rapid increase is expected in the use of granular materials both as fertilizers and as carriers for pesticide chemicals. Granular formulations are not widely used in large-scale locust or vector control programmes because of the added formulation cost and larger bulk, which increase application costs as well. Here, LV and ULV spray programmes seem to be favoured, although these do cause widespread contamination of the treated areas, and public pressure for reduced and more precise use will increase.

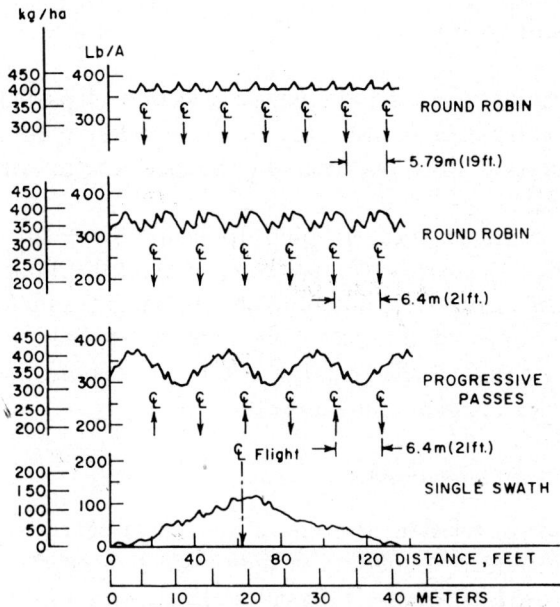

FIGURE 47. Graphs of a Stearman aircraft with single spinning disk spreading granular fertilizer. From bottom to top: single swath; progressive passes (see Fig. 48) on a 6.4 m (21 ft) swath; round robin (see Fig. 48) on a 6.4 m (21 ft) swath; and round robin on a 5.8 m (19 ft) swath.

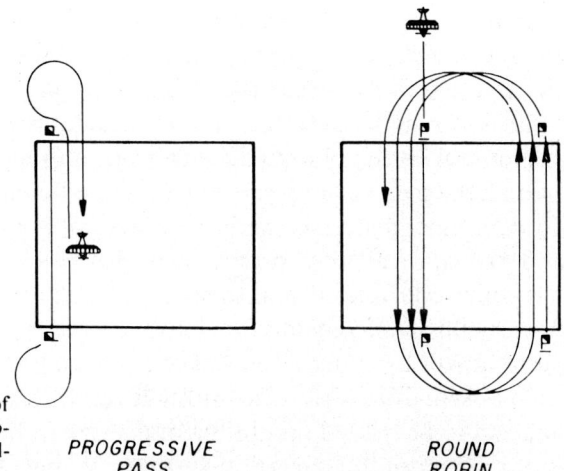

FIGURE 48. Two ways of covering a field: (a) progressive passes; (b) round-robin system.

Liquid formulations

Liquid formulations offer the widest selection of particle size and application volume, and also require the greatest need for understanding and skill in application on the part of the aircraft pilot and operator.

The basic characteristics of liquid applications are (*a*) the formulation's physical properties; (*b*) the atomization of the liquid to a given drop-size range and the limitation of size range; and (*c*) the role played by weather conditions at the time of application, not only in relation to chemical effectiveness, but also to deposit and coverage of the particles on plants and insects.

Particle size and effectiveness

Particle size in relation to spray effectiveness depends in the first place on the mode (or modes) of action on the target pest. With nonsystemic herbicides and fungicides this is primarily a matter of foliage contact and coverage. For insects, however, at least three modes, or entries, are possible: (1) picking up the chemical from plant surfaces; (2) ingestion of the chemical by chewing or sucking; and (3) chemical fumigation or entry of gas molecules into the insect's breathing system (Watkins and Norton, 1955). The old arsenical insecticides were almost wholly dependent on ingestion, while the chlorinated hydrocarbons, DDT, and others appear to be effective as both contact and ingestion agents. In many cases it is difficult to isolate the action, since contact with the insect body may cause the insect to remove particles from the body and ingest them. The fumigation action (Hoffman and Lundquist) appears to be more ccmmon with the phosphate chemicals, although the vapour pressure of malathion, for example, is not all that different from DDT (Table 8). Of course, all materials can be vaporized at high enough temperatures, and the co-distillation process can also effect vaporization, as when a readily vaporized solvent such as acetone (Table 8) is used to dissolve a normally solid chemical. In this case rapid evaporation of the acetone will carry molecules of the pesticide chemical off with it.

The actual particle size most effective for contact or ingestion appears to be related first to the insect size — for example, the smaller instar of mosquito larvae are unable to ingest particles of over 25 μm, and smaller particles are generally more active in the gut of all insects as well as in external contact with the insect (Smith and Goodhue, 1942). Compromise is necessary, however, when materials are distributed in the air, as difficulties of deposit arise when particles under 100 μm are used for aircraft application. Other factors, such as the particle shape or the liquid carrier or formulation, may affect the chemical activity and action with regard to specific insects or even to physical retention on surfaces (Hadaway *et al.*, 1970).

As insect size increases, the contact efficiency or the optimum particle size increases — locusts flying through the air collect from 60 μm up to 280 μm drops more efficiently than they do the smaller sizes (MacQuaig, 1962). In other instances, such as tsetse fly control, where forest canopy penetration is essential, drop sizes of 25-50 μm appear to be most effective, although this is not easily obtained with any of the atomizers available and requires a special operation (Burnett, 1962; Lee *et al.*, 1969). This technique of using a highly vaporizable spray fraction produces smaller drops by evaporation of the emitted spray; however, this limits the number of drops to those initially produced, and with evaporation and loss of the very smallest, even fewer drops remain to provide coverage from the total volume of spray used.

The direct impingement of drops of the chemical toxicant on the insect has always been seen to offer the most efficient prospect for insect vector control (Himel, 1969). Control of adult insects as agricultural, forestry, or vector problems sometimes requires techniques for rapid and effective knockdown with the hope of some residual effects. However, residual control comes from depositing chemicals on either water or ground as breeding sites for larvae or on plant foliage where insects come to rest. Adulticiding techniques using aerosols may give deposits on water and ground, but seldom on foliage. Extensive efforts have been directed toward aerosoling for adult vector control (Mount, 1970), which is, however, extremely complex, highly dependent on weather, and frequently less than successful when applied by aircraft.

There has been considerable research directed toward determining the relationships between drop size, meteorological conditions, and direct-contact insect control by aerial application (see Latta, *et al.*, 1947; Johnstone, Winsche, and Smith, 1949); however, the final particle-size composition adopted for any aircraft use will have to be a compromise between the optimum for maximum-recovery swath width, penetration of vegetation, and toxicity to pests and minimum damage to food, fish and other wildlife organisms (Kruse, Hess, and Ludvik, 1949).

The collection of drops on flying insects corresponds to the Sell relationship as expressed in the following formula (see Latta, Randall et al., 1947):

$$E = D^2 V/S$$

The collection efficiency (E) is shown to increase rapidly with drop size (D^2) and less rapidly with increase in the relative velocity (V) of the insect or of the drop to insect, and decreases as the insect or object size (S) increases. Figure 49 shows collection efficiency of rods 3 and 13 mm (1/8 and 1/2 in) in diameter for 10, 20, 40, and 100 μm drops in increasing airstream velocities. Another aerodynamic collection function is object shape, such as indicated by the high collection efficiency of flat plates, even on the side away from the airstream, in comparison to spheres (Brooks, 1947).

Many observers (e.g., David et al., 1946) have shown that wind-transported drops having a velocity even as low as 1/2 to 3 mi/hr have greatly increased dynamic catch or deposit efficiency. Thus the deposit efficiency of the spray on insects is tremendously enhanced if the insect flies through the spray-laden air. The function of

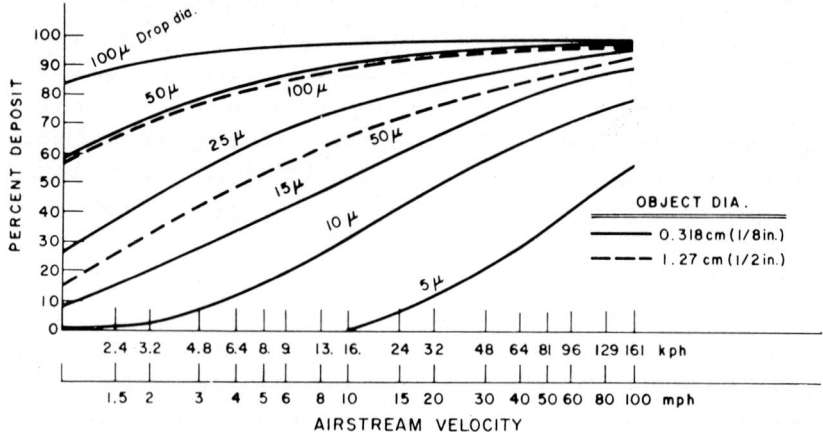

FIGURE 49. Graph of theoretical deposit versus airstream (liquid drop or particle velocity) for several drop sizes and two object sizes.

drop size squared times the drop velocity (D^2V) is used to identify the size and velocity relationship. When D is in microns and miles per hour, a value of 300-1 000 with a maximum D of 100 μm gives a range of values acceptable for adult mosquito contact. It is to be noted that the drop size containing an LD_{100} dose of chemical (a dose capable of killing 100% of the test insects) becomes a limiting factor as regards efficient size (Weidhaas et al., 1970). Therefore, drops of 25 μm for malathion, 20 μm for naled and 17.5 μm for fenthion containing LD_{100} doses for certain mosquitoes would also be the most efficient drop sizes (of technical materials) to use for adulticiding or direct insect contact. These are in the aerosol classification, however, and after release from an aircraft will be rapidly dispersed. Table 7 presents data on the terminal velocities of various drop sizes, the number of drops of a given size for a given volume applied to a flat unit area, and the number of drops per cubic centimetre of air if the drops caught on the flat area are dispersed to a depth of 20 m (65.5 ft). As can be seen, the number of drops available decreases rapidly as the drop diameter increases (as d^3). At 9.35 l/ha (1 gal/acre) applied volume the theoretical number of drops of 100 μm diameter dispersed over one acre to a depth of 20 m (65.5 ft) is about .09 drop/cm^3 of air; at 10 μm this is 90 drops/cm^3, which improves the contact efficiency enormously if the insect actually collects 10 μm drops.

The phenomenal success of coarse aerosols (75-100 μm VMD) of highly toxic materials has been demonstrated for adulticiding mosquitoes and locusts (CIBA, *The principles of waterless spraying*, 1969). For example, when applications for mosquito control at rates of 1/4-5 oz/acre were made, 24-hour kills were found to be very high, although very little residual deposit resulted and control after 24 hours dropped very fast (Glancy et al., 1966). The low dosage of 1/2 oz/acre would place only a theoretical 2.25 drops/in^2, or .0002 drop/cm^3, in an air depth of 20 m. The results indicate the extreme toxicity of the chemicals to adult mosquitoes, but they also show the non-residual character of phosphate materials no matter what drop size (coarse or fine) is used. Applications of phosphate materials in forest areas generally require a much higher volume and dosage of active chemical to obtain results.

Research has shown that formulations having the highest concentration of toxicant in a drop are most effective in transmitting a killing dose to the insect (McGovern, Fales, and Goodhue, 1943). The present trend in the use of highly toxic materials is toward technical concentration, or ULV spraying, using as large an amount of toxicant as can be dissolved in a given volume of liquid. When dilution is needed, such as when the toxicant would be effective in a very small technical dosage, it is sometimes desirable to dilute the toxicant with agricultural spray oils or other materials, such as glycols and glucose. These are relatively non-evaporative, and control of the drop size produced remains at the nozzle. These relatively non-evaporative and low-vapour-pressure materials generally raise the viscosity while lowering the surface tension (Table 8). The net result, however, is larger drops from these formulations — a factor which, of course, must be considered when spray application is to be made (Maas).

Contact with drops or crystals of chemical on plant surfaces, such as grasses, trees, and brush, constitutes another mode of insect contact with chemicals. The residual effects of DDT, and other chlorinated hydrocarbons are attributed to the deposit of the chemicals and contact with the deposit by the resting insect vector. The highest efficiency of recovery for deposits on plant surfaces comes from relatively coarse sprays of 300-450 or even 700-800 μm VMD (Butler, Akesson, and Yates, 1969). Ideally these large drops (above 500 μm VMD) would virtually eliminate any aerosol transport or drift of the spray while depositing 90-95% on the target area. But, as can be seen in Table 7, the number of drops decreases dramatically as size is increased, so in order to maintain coverage with very coarse spray the volume must be raised to 93 l/ha (10 gals/acre) or more. But the use of smaller drop size (and small volumes of liquid) is almost essential if the cost of application for the unit of area covered is to be kept at a minimum. On the opposite end of the spectrum, drops of 10 μm and under do not collect on foliage and grasses, and hence these fine aerosols will not provide any residual action by contact.

Theoretical knowledge and practical experience indicate that, with the present limitations of atomizers, there can be no specific drop size best suited to all the conditions surrounding even a given insect, so a compromise must be reached for each job to be done. Again, the importance of highly toxic chemicals to the ULV concept cannot be overstated. Before resistance build-up, materials such as DDT, malathion, dieldrin, BHC, and parathion all had phenomenal killing power on a wide range of vector, forest, and agricultural insects. The concepts of ULV and the use of fine sprays or coarse aerosols offer a tremendous tool for large-area control programmes. The problems of aerial transport and drift caused by the fine sprays or coarse aerosols (under 100 µm VMD) cannot be ignored, however, and will have to be given increasing consideration in future large-area programmes as well as in agricultural work on a smaller scale (see Table 2).

Control of spray drop size

Controlling spray drop size for precision spray applications appears to be the most logical approach to reducing, if not eliminating, the aerial transport or drift of small spray particles. The problem of chemical pesticide losses to the general environment, where damage to fish and wildlife can take place, is of serious concern the world over. Chemical loss can take place (*a*) by direct aerial loss at the time of application; (*b*) by direct application losses to soil under sprayed crops; (*c*) by transfer from crop plants to soil and thence to ground water (not large with low water solubility); (*d*) by losses to water run-off or to wind transport of plant materials and soil particles; and (*e*) by losses of pesticides carried on raw produce processed in central canning, freezing, or fresh-food plants, in which case the pesticide chemicals are peeled or removed with outer leaves or are washed off and become contaminants of sewage systems, from which they flow into the waterways and finally into the oceans (Akesson, Wilce, and Yates, 1970).

Application techniques for pesticide chemicals can greatly affect the losses at the time of application (Finkelstein, ed., 1969) and, to a lesser extent, the losses to the soil by runoff from crops. In this regard the technique of run-off spraying of plants, which causes high soil contamination, has never been a problem of aircraft application, in which 1/10 to 1/25 the volume of ground-rigs are customarily used.

In addition to the environmental contamination, and probably of much more direct concern to food producers, are the damages caused to agricultural crops in the area where the pesticide chemicals are used. There are four very practical and direct reasons why drift control of pesticide chemicals is of the first order of importance to agriculture today (Akesson, Wilce, and Yates, 1971).

1. Drift losses varying from a few percent to over half the applied material, depending primarily on particle size, reduce the amounts of chemical available at the crop site to control pests.
2. Hazards to nearby workers and hazardous working conditions for pilots, loaders, and others concerned with application are of great concern, especially with the phosphate and carbonate types of chemicals.
3. Better drift-control equipment can reduce the amount of time lost due to either windy weather or to the opposite and, even more hazardous, quiet temperature-inversion conditions, which can cause local damage by a build-up of airborne pesticide concentration in the absence of air ventilation.
4. Lastly, drift control can reduce the potential transport of herbicides, which can damage nearby susceptible crops, or of other pesticides, which can cause over tolerance residue on nearby crops. Even more subtle is the damage caused to the local ecology or agricultural insect ecosystem, where even very low drift levels of pesticides can reduce or eliminate predator and parasite insects. The loss of these can permit economic insect populations to expand very rapidly, frequently beyond the control of further chemical applications.

Formulations and drop-size control techniques for reducing drift are the following:

1. Phase-crowded emulsions, generally invert types (see Table 8) with the oil phase continuous and the water broken up as small particles. These are widely used with phenoxy herbicides for controlling trees and brushy species along power-line rights of way, ditches, and roadsides. To reduce the numbers of small drift-prone drops, the average or VMD has to be made very large. This greatly reduces coverage, so larger volumes (up to 200 l/ha, or 21 gals/acre) are applied in order to provide a sufficient number of drops for coverage with the 2 000-5 000 μm VMD sprays.
2. Viscosity additives such as Dacogin, Vistik, and Norbak (Butler, Akesson, and Yates, 1969), which make the liquids very viscous and thus produce large drop sizes in a manner similar to the

invert emulsions. The use of these is also similar to that of inverts.

3. Mechanically produced large drops, such as from the jet nozzle shown on page 58 (Figure 24b, below). The drop size can be as high as 1 500-2 000 μm VMD at the speed of 80 km/hr (50 mi/hr), but decreases to 800-900 μm VMD at 161 km/hr (100 mi/hr). This nozzle is widely used for applying phenoxy herbicides in hazardous areas where damage potential is high. These drops are also poor from the point of view of coverage and a minimum volume of 93.5 l/ha (10 gals/acre) is required to ensure weed control.

4. Small-drop cutoff nozzles, such as the controlled-breakup jet produced in the low-turbulence area behind an airfoil section as shown in Figure 24b (above). However, this nozzle cannot be operated at over 97 km/hr (60 mi/hr) and produces 800-1 000 μm drops, but almost none below this size. Again, the very coarse drops do not provide good coverage, and until devices that produce uniform, smaller drops are developed their use will be limited to certain systemic herbicides, primarily for brush control (Wilce et al., 1974).

5. Other spray additive materials, such as foam-producing and viscosity additives, have been introduced and show promise of limiting drift. Whereas foam additives do not of themselves limit drift, the use of foam to extend plant coverage with large spray drops may offer desirable results with certain chemicals. Visco-elastic agents appear to produce a long-chain molecular structure in a water-base spray that tends to reduce the number of small drops formed and thus reduces drift.

6. Systemic or translocated materials which can be applied to soil as large, granular particles or to plant surfaces as large spray drops also offer a basic means of using large, low-drift particles without a loss of chemical efficacy owing to poor plant coverage.

7. Granular materials which have been sieved to remove driftable particles, usually below 100 μm, offer a highly practical means of controlling drift where systemic granular materials can be used. Microgranules composed of particles sieved to between 50 and 100 μm are also being used in Japan, where they have proved

particularly successful for application to paddy rice. On the other hand, the limited adherence to and spreading over the plant surfaces are not likely to encourage the use of granules with the nonsystemic type of chemicals.

The relative drop sizes produced by a range of nozzles is shown in Figure 50, which plots drop diameter (vertical scale on the right) and a percentage summation of drop size (below). This is a logarithmic probability graph, which means that the drop distribution at the 50% line is the VMD, or volume median diameter. For example, spray from a D8 jet nozzle directed with an airstream of 105 km/hr (65 mi/hr) at 2.1 kg/cm² (30 lbf/in²) liquid pressure and with the addition of Norbak is very coarse at around 8 000 μm VMD. The second graph down is for a D8 jet with 2.5% oil in an oil/water emulsion at the same velocity and pressure as in the first graph. Drop size decreases to a little over 1 000 μm. In the next graph an increase to 161 km/hr (100 mi/hr) decreases the drop size to 850 μm. The fourth and fifth graphs are almost identical at 420 μm VMD, with a D7-45 at 3.5 kg/cm² (50 lbf/in²), but no airstream, and with a D64-6 in a 161 km/hr (100 mi/hr) airstream (Coutts and Yates, 1968). This illustrates the effect of air flow on drop breakup, as well as the effect of a 46 versus a 45 whirlplate. The sixth graph is for a D-646 directed downward to produce a drop size of 290 μm VMD. The next two graphs are for fine sprays: the first a small spinner and the second an 80005 fan-type nozzle directed 45 degrees downward. Both were operated in a 161 km/hr (100 mi/hr) airstream, and the liquid pressure and fluid used were the same. The drop size is a little above 100 μm VMD. The last graph is for a twin-fluid (air and water) nozzle like the one shown on the far right in Figure 24a. This nozzle, with 0.35 kg/cm² (5 lbf/in²) air and water pressure and no airstream, produced drops of around 15 μm VMD. Adding an airstream would not be likely to affect the drop size any further, since the air velocity required to break up even a 100 μm drop would be around 200 km/hr (120 mi/hr), as can be seen in Table 10.

Aerial transport or drift

The amounts of downwind drift produced by various nozzles and fallout from distances of 1.6 km (1 mi) or more are shown in Figures 51 and 52 (see Yates, Akesson, and Coutts, 1967; Coutts and Yates,

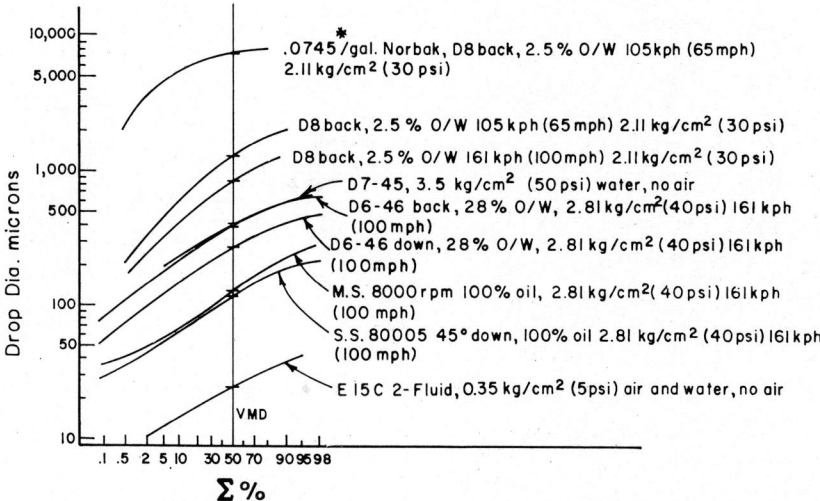

FIGURE 50. Graphic presentation of observed data showing logarithmic (vertical drop-size scale) versus statistical probability (lower scale) of drop-size distribution and the 50% or VMD (volume median diameter) line for several types of atomizers.

1968; Yates and Akesson, 1966; Akesson and Yates, 1964). The data shown are collected by placing replicated 18 × 46 cm (6 × 18 in) plastic sheets at several points downwind from an established and controlled single aircraft path. The aircraft is flown on the given path for eight to ten passes, and the total fallout collected on the sheets is washed off and evaluated for each station. A precise amount of tracer material (fluorescent dye or metallic salts) is added to the spray mix, and the amount on each collection sheet is determined by analysis with a fluorometer. These instruments have a sensitivity to 5 or 6 parts per billion (per thousand million), comparable to that of liquid-gas chromatography analysis of pesticide chemicals.

Figure 51 shows the downwind fallout from aircraft application as indicated for each test. The topmost graph (for greatest drift) shows an 80015 fan nozzle directed 90 degrees to the 145-161 km/hr (90 100 mi/hr) airstream. This and all tests shown were carried out at 2.8 kg/cm² (40 lbf/in²) liquid pressure. The drop size produced is around 180 μm VMD; the liquid was a low-volatility petroleum oil. The drift levels are high: 0.0012 μg/m² (1 μg/ft²) at 1 500 m (5 000 ft) and over 0.12 μg/m² (100 μg/ft²) at 150 m (500 ft). This is the type of nozzle that is customarily used for ULV or fine-spray operations in large-area programmes. The next graph down is for a D6-45 hollow-

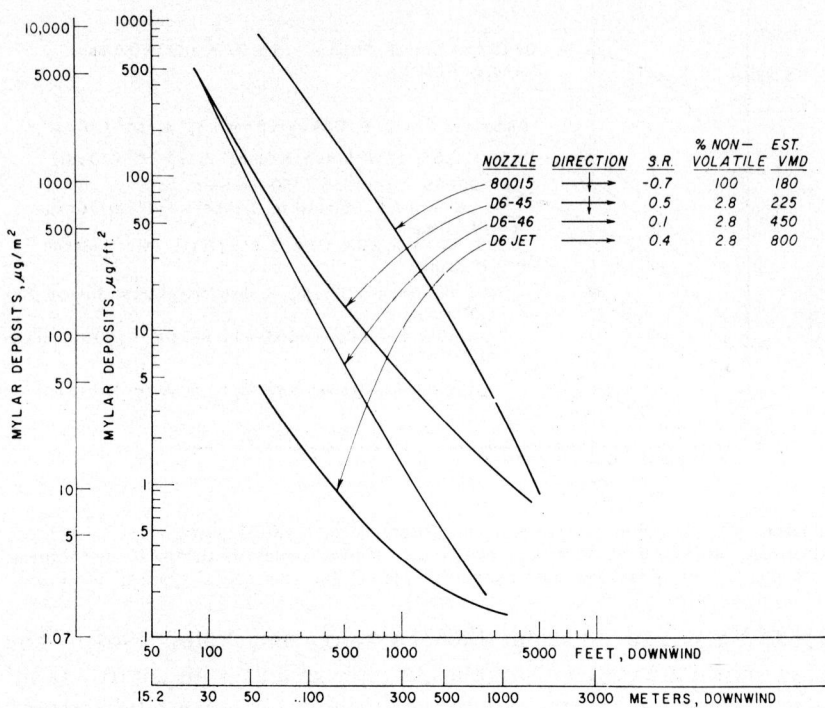

FIGURE 51. Downwind air-transport levels of actual field-measured spray drift. Vertical scale (logarithm) is ground-level deposit in micrograms per square metre (ft²), and lower scale (logarithm) shows downwind distance in metres (ft). Data for four different atomization or drop-size distributions (Fig. 50) are shown.

cone nozzle, also directed at 90 degrees and spraying a 2.8% oil-content oil/water emulsion. The drop size of 225 μm VMD decreases drift significantly, almost tenfold at 150 m (500 ft), without much difference at 1 500 m (5 000 ft) downwind due to the settling out of the larger droplets. The third graph is for a D6-46 nozzle directed with the airstream and using the emulsion formulation, which produces a 450 μm VMD drop size. Again a significant decrease in drift is obtained — somewhat less at 150 m (500 ft), but on the order of a tenfold reduction at 1 500 m (5 000 ft). The D6 jet directed back gives the lowest drift of this series, showing a significant drop of about ten times at 150 m (500 ft) and less drift at 915-1 500 m (3 000-5 000 ft) downwind.

Still other examples of drift control are shown in Figure 52 (Akesson, Wilce, and Yates, 1971). Here, the D6-46 graph is again shown at

APPLICATION TECHNIQUES: PARTICLE BEHAVIOUR

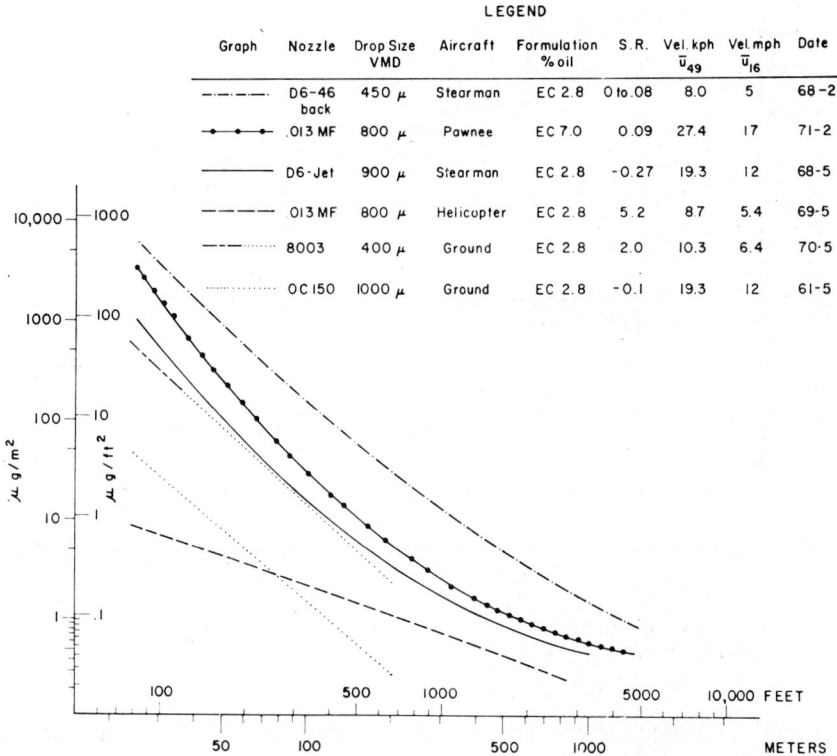

FIGURE 52. Downwind air-transport and ground-level residues from several aircraft and ground-operated spray machines. Different levels of drift are related primarily to drop size of the spray distribution.

the top, as in Figure 50. The next graph down is for the low-turbulence (sharp drop cutoff) nozzle on a fixed-wing aircraft at 145 km/hr (90 mi/hr). Because the air velocity is above the critical value for the 800-1 000 μm drops formed with this nozzle, further breakup occurs, and the drift is of the same order as in the next graph down, for the D6 jet directed with the airstream. The bottom graph shows the Microfoil (® patented) low-turbulence nozzle with very low downwind drift residue due to the sharp cutoff of drops below the 100 μm size.

For comparison, graphs for two ground-rig runs are also shown. The one just below the D6 jet is for 8003 fan nozzles with 400 μm VMD drops, and 2.8% EC formulation, and the lower one with

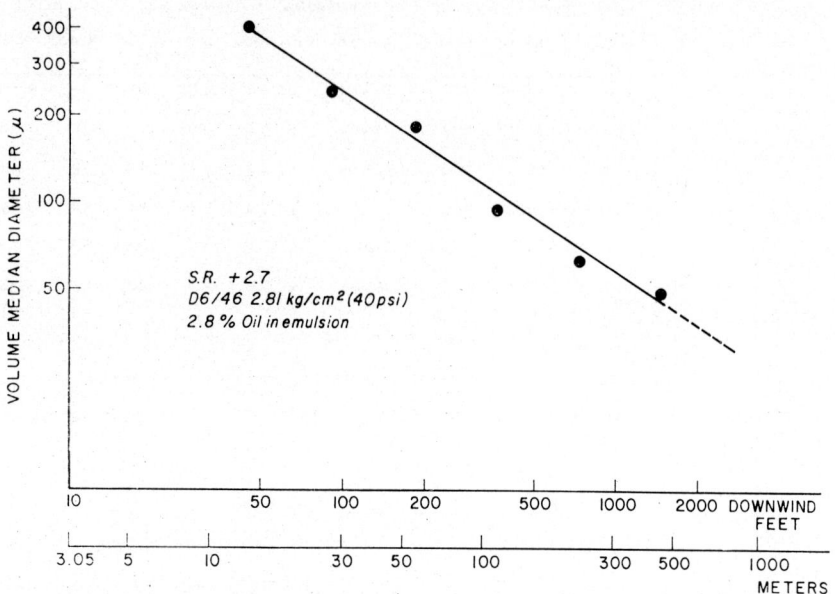

FIGURE 53. Downwind drop-size recovery from field data taken with a water-base spray, initially at 450 μm VMD.

1 000 μm VMD drops, using a very coarse OC-150 fan-type nozzle. Thus it is shown that downwind drift can be a serious problem with ground rigs as well as aircraft, but the potential is of course much lower. With the same drop-size distribution and weather conditions, the drift becomes much the same in both cases (Akesson and Yates, 1964). Other factors such as weather, aircraft application height, and the effect of variations in spray formulation on the evaporation rate will also influence both the deposit characteristics and the amount lost to drift.

Drift losses

The aerial chemical losses from a treated field are not easily accounted for by collection techniques, which need to include both fallout and airborne portions of the spray. While Stokes Law data are often used (Table 7) to show the fallout rate of drops from still air, this seldom, if ever, exists; small drops under 25 μm may never deposit anywhere in the area being treated. From 100 μm on up, the Stokes Law data and the HU product, widely used by the locust-control people, can be satisfactorily utilized to predict downwind deposit of a given size drop or drop range (Sayer, 1966). Figure 53 shows the VMD drop-size range found at various distances downwind from the aircraft flight path (Coutts and Yates, 1968). The initial size in the swath was 420 μm VMD, and at 60 m (200 ft) the remaining size was 200 μm. At 152 m (500 ft) the size had decreased to 100 μm, and at about 365 m (1 200 ft) the size was still 50 μm VMD. Other data for as far out as 10 miles have shown particles as large as 25 μm in the air.

From this type of information it would appear that there is little object in putting an aerosol into the air with any sort of deposit expectations. In fact, when the weather conditions are highly stable, an aerosol introduced at a given level will move at this level with little change in vertical height and will not be of any value, except as a short-duration adulticiding measure. It would also seem logical that if this small-drop fraction of under 50 μm could be eliminated, either before they are produced or before they are released, the principal component of drift loss would also be eliminated.

Recoveries of spray have been made by many observers. Yeo (1961), using Micronair spinners, showed the general increase in the percentage recovery (of low-volatility sprays) as VMD increased. Thus at 134 μm VMD the recovery by volume was 75%, while at 73 μm VMD it fell to 49%. Table 11 shows the general size ranges and their uses, from adulticiding space aerosols to coarse total-confinement sprays. The approximate VMD diameter is shown in column 2, and the estimated percent recovery is given in column 3. Thus, while 99% recovery can be expected from the Microfoil (® patented) coarse-spray boom, the recovery diminishes rapidly; at the coarse aerosol range of around 100 μm the recovery under low wind conditions within a 305 m (1 000 ft) distance would not be over 25%.

Figure 60 shows the recovery in a swath (several passes) from a D6 46 nozzle system (450 μm VMD) on a Stearman biplane. The basic pattern shows about a 15 m (50 ft) swath with less than 65% of the applied chemical being recovered. If the collection is extended to 25 m (82 ft), the recovery is 70%. Another 8% is added if recovery is carried to 50 m (160 ft), and about 4.5% more out to 200 m (660 ft). The total collected in 400 m (1 315 ft) is close to 84%. Thus, by subtraction, 16% of the applied material goes beyond 400 m (1 315 ft) downwind. At the edge of the field this, of course, would be a loss to the field being treated or a drift loss that could cause problems in adjacent fields. But as the aircraft working against the wind applies farther into the field, the losses decrease since they collect instead in the treated field. As noted in Table 11, however, these losses can be much higher as spray atomization is increased or drops are made smaller.

Spray swath patterns

The swath patterns laid down by the aircraft sprayer are also affected by the drop-size range, as well as other factors arising from the relationship of boom to wing or rotor length to flight, altitude, and in some respects to aircraft size (Yates and Akesson, 1966). The basic aerodynamic vortex pattern created by the aircraft controls the initial transport of the spray drops. Both fixed- and rotary-wing aircraft produce wing- or rotor-tip vortices (Figs. 4 and 5), while the propeller vortex on the fixed-wing aircraft adds a third factor, which shows up particularly with large engine aircraft as a displacement of spray or even granulars at the centre of the swath.

Figure 54 shows the pattern of distribution from a fixed-wing aircraft, a Stearman biplane with 450 hp engine, using a 10.3 m (34 ft) wing-length boom and a 12.2 m (40 ft) boom, which extended beyond the tips of the wing. Three test runs are shown in the drawing. The first is the dotted line designated test E, where D7 jet nozzles without whirlplates were used, and shows the typical cutoff at the edges of the pattern that occurs with this coarse spray of 800-1 000 μm VMD drops. In the second test, F, the longer boom was used with the D7 jets, and the swath was extended about proportional to the extra 3 m (10 ft) of the boom, or from 12.2 m (40 ft) to about 15.2 m (50 ft). In the third test, G, with a considerably finer spray of about 350-400 μm VMD, as opposed to the jet-spray size of around 800 μm VMD, the finer spray is seen to add to the swath width, now raising it to about 16.8 m (55 ft).

The helicopter shows similar spray characteristics, with rotor-tip vortices affecting the dispersal pattern and in some instances causing spray peaks as a result of vortex disturbances. Figure 55 shows a Bell helicopter run with D6 jets for a very coarse spray of 800-1 000 μm VMD and D6-46 jets for a drop size of about 500 μm VMD. The

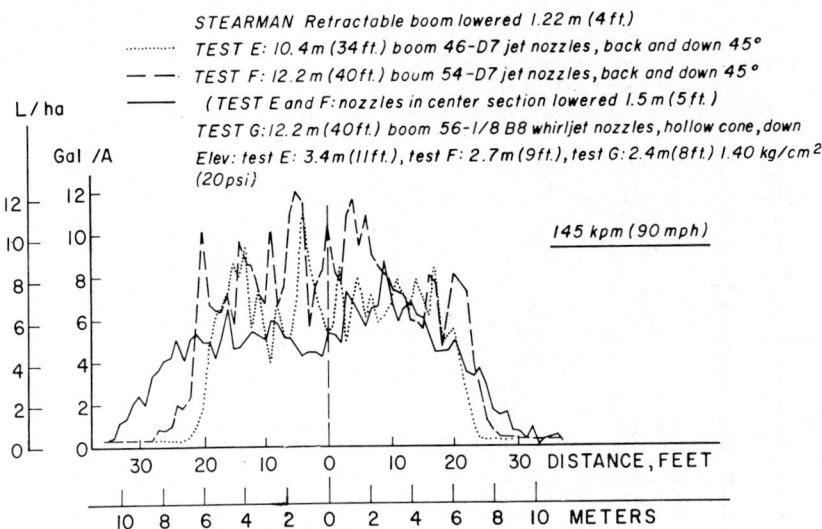

FIGURE 54. Several single-swath graphs of different spray-drop-size systems used on a Stearman aircraft.

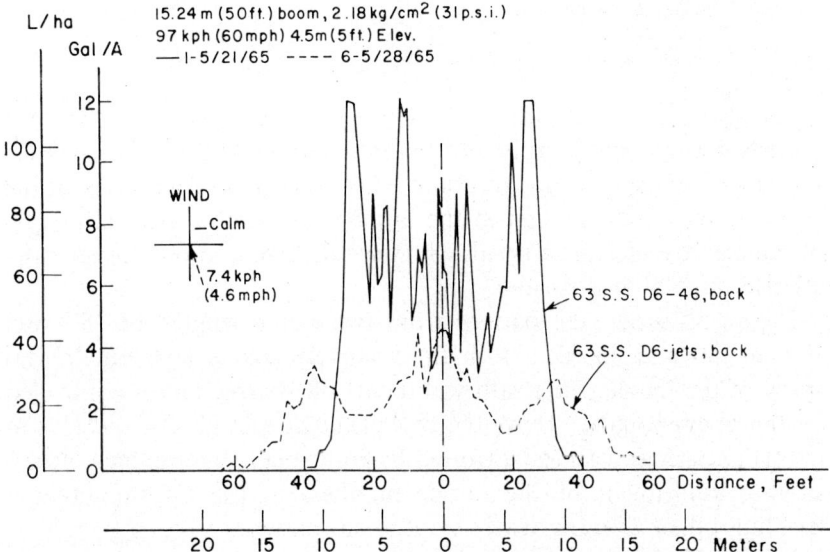

FIGURE 55. Two graphs of single-swath distributions of (lower graph) coarse-spray (900 μm VMD) and (upper graph) medium-spray (500 μm VMD) applications from a Bell 47AG helicopter with 15 m (50 ft) boom.

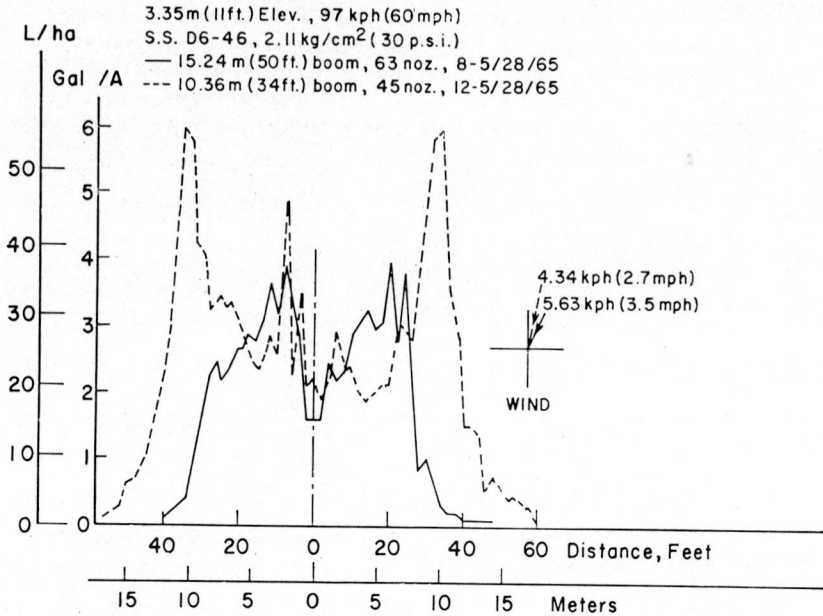

FIGURE 56. Graph of medium-spray (500 μm VMD) application with two boom lengths on a Bell 47AG helicopter.

D6 jets cut off very sharply at a boom length of 15.2 m (50 ft), while the D6-46 jets spread out to about 33.6 m (80 ft). The levels of the two runs are different due to the number of nozzles used, although, of course, the increased swath width would give a lower dosage rate, all else remaining the same.

Figure 56 shows the pattern from two boom lengths of 15.2 and 10.6 m (50 and 35 ft). With the relatively coarse (500 μm VMD) spray of the D6-46, the swath width cuts off sharply and is controlled by the boom length. Here, the two peaks at the sides of the 15.2 m (50 ft) pattern appear to be caused by rotor-blade interference, necessitating adjustment of the nozzle numbers at the tip to obtain a levelling off or a more nearly trapezoidal pattern.

Figure 57 shows the spray pattern that can be obtained with a Bell helicopter equipped with a 15.24 m (50 ft) boom and using 63 D5-45 nozzles directed downward or at 90 degrees to the 93.4 km/hr

APPLICATION TECHNIQUES: PARTICLE BEHAVIOUR 101

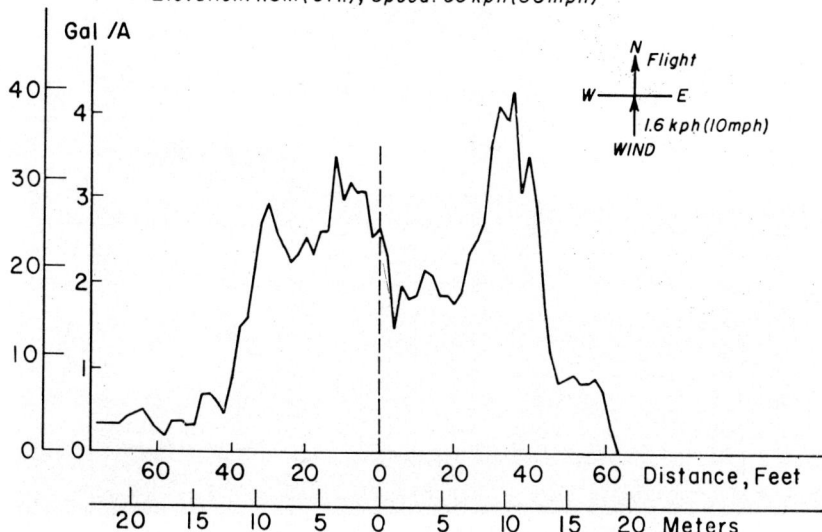

FIGURE 57. Fine-spray distribution (350 μm VMD) from a 15 m (50 ft) boom on a Bell 47AG helicopter.

Figure 58a. Multiple-orifice nozzle, showing spray pattern of inverted (water in oil) emulsion.

FIGURE 58b. Bell 47AG helicopter spraying with a narrow-range drop size (800 μm VMD) microfoil, using a low-turbulence nozzle system.

FIGURE 59. Closed transfer system for handling hazardous chemicals. Concentrate liquid is drawn from the storage barrel (centre) into the mix machine (right). Mixed material is transferred to the aircraft by a tank-bottom loading system.

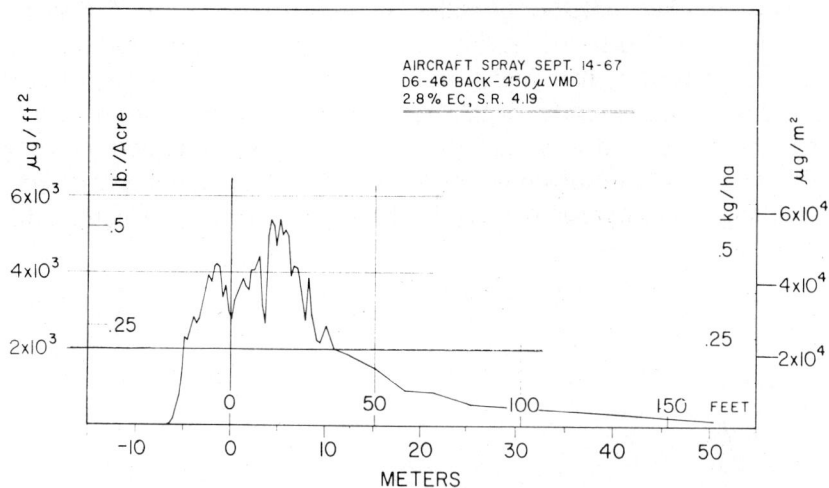

FIGURE 60. Graph of deposited spray residue recovery from medium (450 μm VMD) spray applied by a Stearman aircraft.

(58 mi/hr) airstream. The drop size with the water-base (2.8% oil-water emulsion) spray is estimated at 250 μm VMD, and the usable (flagged) swath is about 21.3 m (70 ft). Thus it is seen that as the spray drops are made smaller, the swath width increases significantly, well beyond the 11.3 m (37 ft) rotor width. Of course, the finer spray also produces greater drift losses.

FIGURE 61. Field operation, showing a Hughes helicopter landing on a vehicle-mounted platform for rapid field-loading procedure. *Photo Hughes Tool Co., U.S.A.*

This graphic material provides an approximate basis for swath widths to be taken by various settings of spray atomization. In practice, however, it is highly desirable to check the swath patterns for the specific aircraft, nozzle type and placement, flow rate, and material to be used so as to ensure that a reasonable pattern is being dispersed, without obvious skips or peaks that could cause field streaking or resurgence of pests in the skipped area (see Chapter 10).

9. METEOROLOGICAL FACTORS RELATING TO AIRCRAFT APPLICATION

Probably the most universally involved and yet the least understood of the several factors dominating the application of chemicals by aircraft is the role of meteorology, particularly local microweather and its effect on the dispersion, diffusion, and deposit of released materials and, ultimately, on efficient application (Riley and Giles, 1955; Lomas, Frankel, and Hirsch, 1954). It is important to note that the effect of microweather becomes the single most important determinant of the ultimate fate of an aerosol or of particles under 50 μm in size (Johnstone, 1950). The present trend in volume reduction is to the technical level, which for many materials could be 73 ml/ha, or an ounce or less per acre. This, in turn, favours the use of aerosol-size drops. Atomization can be in the coarse aerosol range of 50-75 μm VMD for 1.2 l/ha (1 pt/acre), and should go down to the mid-aerosol range of 25-30 μm VMD for 100 ml/ha (1.4 oz/acre) or less, in order to obtain sufficient coverage or air concentration to be effective.

Unlike the atomization or formulation factor, weather cannot be controlled. The applicator must "live with" the weather and try to confine his application to those times when wind velocity, direction, and gradient, ambient temperature and temperature gradient, and relative humidity are favourable. Operationally this can be frustrating to the commercial applicator, who must try to accomplish the job to be done, frequently in spite of less favourable weather.

While the most efficient drop size for depositing particles on insects in a wind tunnel can be ascertained, application in the field must deliver the toxic material to the area of the insect and in such form that it is available for effective control. The factors pertaining to applications and their interrelation with weather make this one of the most complex and frequently unsuccessful techniques.

Wind velocity and direction

Wind velocity and direction are the most obvious of the weather parameters. Aerosols are by definition small airborne particles and are almost entirely at the mercy of the air motion and turbulence in effect at the time of application (Johnstone, 1950). Turbulent mixing will transport fine-particle sprays over considerable distances to give space dosage, or it will permit different-sized particles to fall out of the aerosol cloud and distribute over an area or swath. The basic instrument for wind velocity is the cup anemometer, which may be coupled to recording instruments for rapid read-out or information storage. Smaller, less expensive hand-held pressure anemometers can also be used with a lesser degree of accuracy.

Wind velocity is generally lower close to the ground and increases with height, depending upon the local geography of mountains and valleys, as well as upon the barometric, or atmospheric, pressure differences. The steepness of the velocity gradient, or the ratio of overhead air velocity to that closer to the ground, provides information on the rapidity of vertical mixing or diffusion. For example, the velocity ratio U_1/U_2, where U_1 is taken at 10 m (33 ft) and U_2 at 1 m (3.3 ft) is greatest under temperature-inversion conditions, when the air is highly stable and no vertical mixing is taking place. Under temperature-lapse conditions the velocity ratio would be less due to vertical mixing.

Vertical temperature gradient

The temperature gradient as observed from measurements at two different heights (such as 10 m and 1 m in the above example) will indicate the degree of vertical mixing taking place. Sensors for measuring temperature difference between heights of a few meters must be sensitive to at least 0.1°, although the temperature gradient from ground level to 500 m (1 640 ft) or more can frequently be observed with a simple thermometer placed outside the cockpit of an aircraft flown to this height. The temperature change with height, or temperature gradient, is one of the most critical factors that control air turbulence and thereby vertical mixing. The normal relationship of pressure, temperature, and density of the air follows the laws of gases. The air near the ground is normally warmer than that above due to adiabatic cooling. Because of reduced pressure with an increase in height, the change in temperature going upward will be at the rate of 1°C/100 m (5.4°F/1 000 ft) of elevation, which is called the normal adiabatic lapse rate. Thus if a parcel of air is moved from elevation Z_1 to Z_2 without adding or removing heat, the change in pressure of the parcel will cause its temperature at the higher level to be less (Pasquill, 1962, Chap. 2). This is called a normal stable or neutral condition, and the air tends to maintain this stability even with low wind velocity. When the incoming heat from the sun causes the ground to warm up, the air also becomes warmer and literally starts to boil upward, causing turbulent mixing in what are called superadiabatic conditions. If, as in the evening and early morning hours, the air at ground level is cooled by the radiation of heat from the ground to the sky, or by the flow of a warm air mass above or a cold air mass below, or by high evaporation of water from transpiring crops or water surfaces, then the ground-level air becomes cooler than the air above and a temperature inversion exists. The air mass, even with considerable wind, can be highly stable with little or no vertical mixing. Maximum temperature inversions usually occur when a high daytime ground temperature is followed by maximum cooling brought on by radiation to the cold sky, which is further aided by evaporation cooling in the case of croplands. Such cold-sky radiation cooling produces an inversion which starts in early evening and may continue through the night, reaching a peak in the early morning and continuing until the sun warms the land. An inversion can be caused by air movement, either by drainage or subsidence from cooling mountain slopes or by cold-air intrusion from nearby oceans, in which case the daily pattern can vary and a steep inversion gradient can be formed in the early evening, which may drop as the intruding wind speed drops during the night. By morning a radiation inversion may continue to have a strong effect even though the wind may have made a 180 degree reversal and cold ocean air is no longer coming in (Schultz, Akesson, and Yates, 1961).

Relative humidity is an indicator of the atmospheric water content as related to the dry-bulb (usual atmospheric measurement) and wet-bulb temperatures. The latter will be affected by the amount of water vapour present in the air. The evaporation of water droplets in air thus becomes a function of the two temperatures as well. Table

12 shows some data for the evaporation rate or lifetime of water drops of different sizes under 30°C dry-bulb and 22.3°C wet-bulb conditions, corresponding to 50% relative humidity. As can be seen, the smaller drops evaporate much more rapidly than larger drops — that is, in proportion to their diameters squared. If the fall rate at terminal velocity is taken for the drop size and lifetime shown, the distance these would fall before being completely evaporated can be calculated. Thus a 200 μm drop will fall 33 m (109 ft) in its lifetime of 56 seconds, while a 50 μm drop lasts about 3.5 seconds and falls only about 9 cm (3.5 in).

Whereas most spray materials are not pure water and would therefore not conform to these calculations, nevertheless the problem of the loss of small drops is very graphically shown and points out the need for keeping the drop size large when using water-base sprays. Of course, this is also the reason for using non-volatile materials for fine sprays or aerosols (Maas, 1971).

Microweather under the Forest Canopy

The microweather under a forest canopy is a unique situation which has been of great interest to groups involved with tsetse fly control (Yeo and Thompson, 1954) and forest insect control. The overall effect of the canopy is to slow the wind velocity very rapidly, at a rate depending on the forest density. An 8 km/hr (5 mi/hr) wind speed over a 15 m (50 ft) canopy may be less than 1.6 km/hr (1 mi/hr) at ground level. Highly complex temperature structures can occur in the forest. The usual daytime lapse and evening inversion occur in a thin canopy, but a complete reversal of this, or daytime inversion and nighttime lapse, can occur in a dense jungle canopy. The heating effects of sun radiation on the top of the canopy can produce a higher temperature than at lower levels, where cooling is taking place. However, the daily pattern of temperature variation will usually occur above the canopy as over open ground.

Aerosols and Microweather

Applying aerosols or dust by aircraft under various weather conditions introduces a basic question as to the ultimate fate or recovery of these particles or their contact with the intended target. The aircraft creates a violent, turbulent wake (Figs. 4 and 5), which is confined and persists under stable air conditions, but expands and disappears under turbulent temperature-lapse conditions. The simple approach to droplet settling is the gravitational fallout, which can be predicted in accordance with Stokes Law. Although this has been widely used to predict downwind swath deposit, its use should be confined to large particles (over 100 μm) and calm, strong inversion conditions only. To apply this even to coarse aerosols (75-125 μm) is to introduce chances for significant error, as Stokes Law, or ballistic fallout theory, precludes any disturbance from air turbulence, thermal or electrostatic effects, or dynamic catch or impingement on plant surfaces because of drop velocity caused by the aircraft wake or by air movement. The rate of fall of particles as large as 50 μm is small compared to velocities generated by turbulent diffusion in the air (Johnstone, Winsche, and Smith, 1949). Table 7 indicates fall velocities for various size drops, but below 100 μm the fall rate, which varies with the atmospheric turbulence, may bear little direct relation to actual deposit on surface and foliage. For coarse sprays, from 200 μm upward, the fall rate is significant compared to the effects of turbulence, and at 500-700 μm a spray can be recovered from ground surfaces at 95-99% of the applied dosage (Table 11).

Materials of high intrinsic toxicity can be very effective in direct air-to-insect contact, but many practical problems arise, including local meteorology, which makes aerosoling by aircraft a very unpredictable technique.

Many observers have shown the benefits of applying drops in a medium to small aerosol range (10-40 μm) from the viewpoint of penetration of forest and crop cover and contact with flying insects (Mount, 1970), but others point to the very low total recoveries, primarily of deposits on plant surfaces, obtained with such aerosols. Aerosols (48-80 μm VMD) produced recoveries of 4-10% on glass slides placed on the ground, while sprays of 60-280 μm VMD gave recoveries of from 40% to as high as 96% (Cutkamp, Hess, and Keener, 1950). Increasing the wind velocity and thereby the drop velocity aids in depositing spray drops (Fig. 49) on vegetation (Sayer, 1959), and sprays of 70 μm VMD have given as high as 57% recovery on vertical foliage when carried by a wind of 8-10 km/hr (5-6 mi/hr). For forest canopy penetration aerosols under 50 μm

VMD will enter the canopy, but little deposit occurs and increasing wind velocity would be of little help; however, increasing wind velocity would not affect penetration by coarse drops, which would tend to settle more nearly vertically through the canopy.

The deposition rate and recovery downwind are of critical concern for aircraft application, but they are not easily determined from our present knowledge of turbulence and diffusion (Slade, 1968). About the best that can be done is to determine the conditions of microweather that exist at the time of any application, and by field-test methods actually try to establish what type of downwind deposit can be expected to occur under the usual weather patterns that prevail in a given air-application area.

The various factors describing local meteorology and affecting spray behaviour have been combined in a function involving wind velocity and temperature gradient. This is termed a stability ratio, or SR, and was introduced as a simplification of the earlier Richardson number (Coutts and Yates, 1968).

$$SR = \frac{T_2 - T_1}{U^2} \times 10^5$$

Here, T_2 and T_1 are temperatures in degrees centigrade taken at 10 m and 2.5 m (33 ft and 8 ft), and U is the wind velocity in centimetres per second at 5 m (16 ft). When SR is positive, the temperature at T_2 is higher (warm air overhead) than T_1, the wind velocity (U) is low and stable, and temperature-inversion conditions exist. When T_1 is larger than T_2 and U is high, turbulent mixing weather prevails and SR is increasingly negative. When SR is near zero, neutral conditions exist — that is, as noted earlier, the normal lapse rate prevails — and mild mixing conditions occur.

Figure 62 graphically shows the effect of stable versus unstable weather on the fallout collection downwind. As can be seen, stable weather (topmost graph) permits less vertical mixing and therefore produces very high fallout levels to 3/4 km (about 1/2 mi) downwind. As SR decreases and then becomes negative, as in the lowermost graph, the downwind fallout is greatly reduced due to rapid mixing and diffusion of the small amount of aerosol-size drops (under 10% by mass) produced by the D6-46 nozzles directed with the airstream.

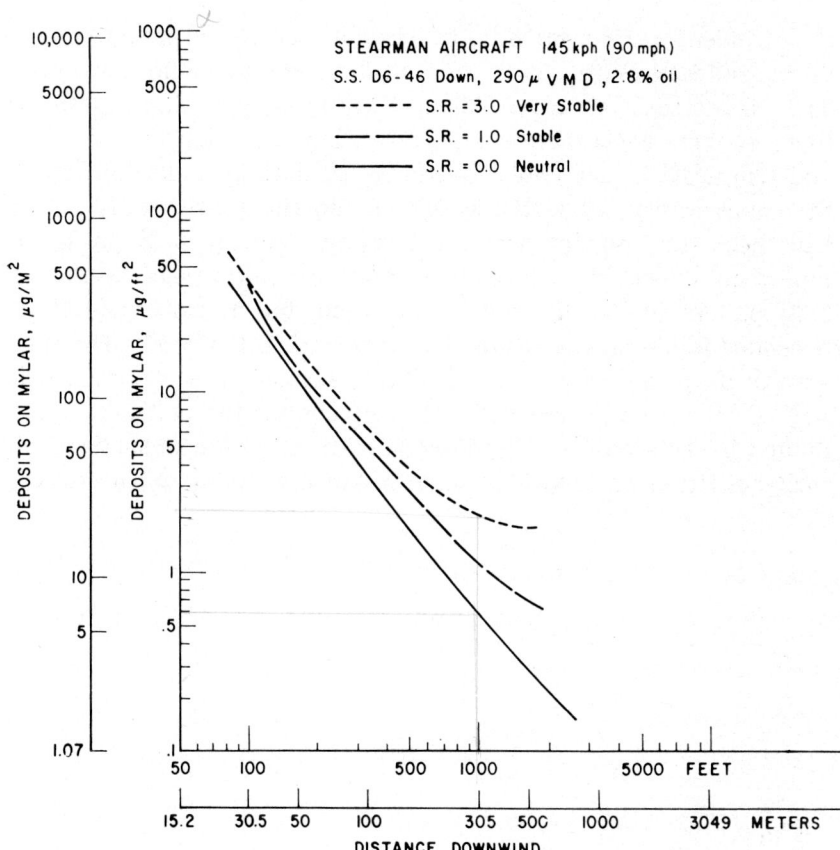

FIGURE 62. Graph of downwind deposit (vertical logarithmic scale) versus downwind distance (logarithmic scale) with medium spray (290 μm VMD) under three types of weather conditions.

At about 150 m (500 ft) the fallout is reduced twentyfold between very stable and very unstable weather conditions.

When aerosols and dusts are applied by aircraft, it is imperative that calm temperature-inversion conditions prevail in order to aid in confining the airborne material to the air basin or local area where it is applied. If the aerosol were applied under a negative SR, or turbulent windy conditions, the airborne material would rapidly move out of the application area and disperse widely at low levels downwind. When coarse sprays having only a small percentage by volume of spray droplets in the airborne or under 50 μm size are used (Fig. 62), the turbulent weather provides for rapid dispersion of these few small drops and causes the least concentration at any point downwind. Thus, if coarse sprays (over 300 μm VMD) can be used, it is desirable to apply these when there is a positive wind of 5-8 km/hr (3-5 mi/hr) and temperatures are cooler overhead, permitting rapid diffusion of the small percent of airborne particles in the coarse spray. Tests have been run applying a coarse spray in wind up to 29 km/hr (18 mi/hr) air velocity, which tends to move the entire swath of the aircraft pattern in the direction of the wind, but actually reduces the deposited fallout in the distances beyond 300 m (1 000 ft). The small aerosol drops are carried for significant distances, depending on the size of the air basin, but with a few exceptions the pesticides are so diluted in the atmosphere and degraded by exposure to sunlight that their collection or deposit at some distance is no cause for concern.

10. LABORATORY AND FIELD ANALYSIS OF AIRCRAFT APPLICATIONS

The ultimate test of any application for control of weed, insect, fungi, and other pests will always be the results obtained under field conditions. For plant nutrition, seeding, and other applications, the field results will generally relate to a uniform and high-yielding crop produced by satisfactory applications. However, many test procedures, including simulated spray applications using dyes instead of toxicants or non-toxic granules and dusts, can be used to provide a measure of the accuracy and uniformity of the job being done in advance of actual pest control or other usual measures of succcessful application.

Spray-drop collection techniques

Spray-particle sizing to determine the range of drop sizes being produced or deposited under the aircraft swath is frequently one of the necessary tests to be made. The techniques for determining drop size in sprays and aerosols cover a wide range, but they can be classified as (*a*) direct means and (*b*) secondary or calibrated methods.

Direct observation can be made of drops allowed to settle in a liquid matrix which supports the drop in its original spherical form. For water drops various dilutions of petroleum jelly and Stoddards solvent (white kerosene) are made up and adjusted so that the water drops penetrate the top layer of fluid, but settle slowly to the bottom of the collection cell or are brought to rest on a heavier layer of fluid in the bottom of the cell (Byass and Courshee, 1951). Similarly, oil drops can be caught in the matrix of a mixture of surfactant (soap) water and cellulose (Gieseke and Mitchell, 1965) or of alcohol and cellulose (Sharp and Bufton, 1963).

Direct-observation sampling can be used for comparison against a secondary measure of size, in order to obtain the stain-size correction or calibration of the secondary measure (Byass and Courshee, 1951). It is then possible to spray simultaneously into the matrix and onto the secondary materials. When counts are made from each, the numbers obtained can be aligned to determine the class size that corresponds to that count, and correction factors for each class can be established. Stain sizes vary with original drop size, material formulation, and, of course, the medium on which they are being caught; it is therefore necessary to determine the secondary stain size for each liquid formulation used.

Other techniques for highly accurate measure of stain size involve micro-pipettes (Byass and Courshee, 1951), which can be used to dispense a given volume into a determinable number of equal-sized drops. Various other devices for producing drops of uniform size are vibrating hypodermic needles and reeds, as well as spinning tops and disks (K.R. May, 1949). These are used to obtain known drop sizes (usually from a matrix or direct observation), and then stain sizes are measured after depositing them on the secondary collection surface. However, the practical lower limit of these devices for uniform drops is around 50 μm, as smaller sizes are unstable and difficult to separate. Hence, for very fine aerosols some type of simultaneous comparison is usually needed.

Another direct observation technique for aerosol-drop counting is to photograph the drops in the air, using a high-speed flash as a light source and a dark-field microscope system (Rathburn and Miserocchi, 1967). Negatives of the film can then be further magnified for counting against some form of calibrated grid or graticule (K.R. May, 1950). There are many systems for capturing drops by causing them to settle in an enclosed space or room (Gieseke and Mitchell, 1965), or by waving a glass slide in a spray cloud (Yeomans, 1950). These include (*a*) magnesium oxide, carbon soot, or fine clay surfaces on glass slides or (*b*) high-gloss printing paper (clay impregnated) or dyed paper, in which case the spray dissolves some of the dye on a red- or black-dyed card and leaves a circular spot (Davis, 1953). Probably the most widely used and most practical is the cleaned glass slide coated with a silicone surfacing, which provides a uniform substrate for holding the drop. Sources of silicone are numerous — the laboratory aerosol spray can for coating glassware probably being the most current and readily available; but other sources, such as auto-body polish and silicone pastes or fluids, can be used as well.

Direct spraying of specially processed photographic film has also been used as well as spraying on photographic printing paper.

Secondary measuring methods have the advantage of a relatively permanent record of drops and thus can easily be used in the field and returned to the laboratory for microscopic analysis. Also, the expansion of the stain size makes it easier to count the drops. Direct observation techniques must be handled quickly to prevent drop evaporation or must be covered in the matrix and then transferred carefully to the microscope for counting. The advantage of the latter, however, is in the direct reading of size, whereas secondary methods require a calibration of stain size to true drop size. An optical correction can be made to determine the original drop size from the flattened drop on a glass slide, but the drop must be placed on the microscope and exposed to evaporation, which would be of concern with volatile liquids. Photographing the drops on the glass slide makes possible an immediate record and reduces evaporation losses to a minimum. In any event, with field-testing techniques for collecting drops it must always be recognized that a wholly biased sample is being taken whether settling or slide waving is used, since in the former the airborne drops do not settle out on ground surfaces and in the latter the waving technique does not impinge the small drops or collects the drops too erratically.

Therefore, drop sampling must be geared to the desired objectives. If field checks of a general nature are desired, then one of the secondary means is satisfactory. If large drops are to be caught (over 500 μm), one of the matrix techniques or clay-coated paper works very well. On the other extreme, for aerosols under 20 μm size, silicone glass or film appears to be best, since a coated secondary method may either allow the drop to bounce off the surface or to be lost in the surface grain. When accurate drop counts and sizing are desired, it is customary to use a settling tower; but it is also possible simply to spray into a closed building, where there is a minimum of air motion and the drops are allowed to settle on collecting surfaces (Coutts and Yates, 1968).

The binocular microscope is usually employed to examine slides and count the drops. A micrometer stage is essential for the plotting of fixed paths across the slide, and an eyepiece graticule or grid, such as the Patterson-Cawood, gives a direct size reference, which can be calibrated with a fixed micrometer slide to the microscope. Various types of above- or below-stage lighting have been used, but the more recent phase-contrast light sources have aided significantly in examining drops, particularly those under 25 µm in diameter. Photomicrographic equipment capable of magnification up to 1 000 times with a dry lens also helps when very small drops of 10 µm and under are being examined.

A great variety of automatic and semiautomatic small-particle counting devices have been developed in recent years. The simplest of these are projectors which put the microscope image on a grid screen for ease in counting. Others split the drop image when adjusted for a given size and may also give it a different colour or appearance on a screen that identifies the particular size for which the splitter is set. The most elaborate counters incorporate a television camera, which is coupled to a microscope and set up to scan the slide either directly or from a photographic negative of the drops. A small screen can be used to observe the camera picture, and a built-in computing system can be used to classify the drops and give averages for the observed drops.

Various mathematical summations can be made from the data obtained on drop sizes and size frequency (Orr, 1966). The simplest is an arithmetic average which sums up the number times the diameter of each size class and divides the total by the total number of drops. A great variety of statistical summations can be made by special weighting of diameter, surface area, volume, sedimentation rate, and other factors (Green and Lane, 1964, p. 229). Number and volume have particular meaning in insecticidal work, and hence are used as number or volume medians. The medians are calculated as the percentage of the total represented by each size class, and when this percent is summed up, the class falling at the 50% point is the median size. The medians for numbers or volumes (or other weightings) fall equally above and below the 50% size.

Graphically the medians can be found most easily from curves plotted on linear versus logarithmic probability graph paper, as shown in Figure 50. The normal, or Gaussian, distribution will plot as a straight line on linear probability paper. However, as most spray distributions are skewed rather than normal, log-probability paper is used because it produces more nearly a straight line (predictable function) when plotting spray-drop distribution. The 50% size or median is the geometric mean, and other statistical functions, such as standard deviation, can be obtained by dividing the size at 84.13% by the 50% size or by dividing the 50% size by the 15.87% size.

To obtain a measure of the distribution, the standard deviation or some other function of statistical analysis can be either plotted or simply columned to show the distribution of drop sizes in the spray-drop spectrum and variations produced by different atomizers.

Dry material collection

Dry materials are presized, and hence elaborate techniques for catching these for size analysis are not needed. However, in distribution studies for evaluation of the amounts that will be caught in an aircraft swath, various coated surfaces are used — oils or silicones on plastic, glass, or metal plates (Yates, 1970) — or trays, pans or baskets for coarse dusts or granules (Brazelton, 1969). Also, collection can be made in cups or pans containing liquids for dusts and, on

FIGURE 63. Plastic-basket technique for determining swath patterns, showing dry material spreading by a Stearman biplane.

Figure 64. Evaluation of spray swath distribution, using plastic plates 7.6 by 15 cm (3 by 6 in) spaced at 30.5 cm (1 ft) intervals.

occasion, for sprays where bioassay techniques are to be used (Gillies, Womeldorf, and Walsh, 1968). The field collection for swath patterns is made by placing one or more (replication) lines of the collectors across the path of the aircraft, and returning the collectors (or their contents emptied into bags or bottles) to the laboratory for analysis. In the case of dry granules it is easiest to count the large particles collected in plastic baskets placed under the aircraft (Fig. 63), or else accurate, rapid balances such as the Mettler type can be used to weigh up the samples. The dispersion and the amounts collected can be plotted as shown in Figures 42 and 47. Dusts may be analysed for active toxicants by gas-liquid chromatography, or various dyes can be incorporated on the dust particles and then stripped in turn from the dust for analysis.

Spray pattern determinations

For studies of spray patterns, colorimetric analysis has been used, or water-soluble salts, such as strontium or manganese, which are dissolved in the spray and washed off the collecting surfaces for analysis on a flame spectrophotometer (Fig. 64; Yates, 1970). Fluorescent dyes, both water- and oil-soluble types, have also proved excellent for very low-level determinations of airborne spray particles. Sensitivity comparable to gas-liquid chromatography can be obtained with fluorescent materials analysed in a light-bridge type of fluorometer (Yates and Akesson, 1963). Since exposure to sunlight causes these materials to degrade very rapidly, this loss must be taken into account. Atomic-absorption spectrophotometry permits equivalent sensitivity if non-degradable salts are used.

Use of bioassay for swath studies

Bioassay techniques are frequently used to measure the success of insect or plant control. A technique for checking mosquito larvicidal control is to place cups of water in the field to be treated, expose the cups to the spray, and then bring them into the laboratory, where mosquito larvae are introduced and kill times are noted. Figure 65 shows data collected from such a bioassay technique (Gillies, Womeldorf, and Walsh, 1968). This was taken in the field during an operational run using a fixed-wing aircraft and spraying a larvicide chemical. Cups were placed at two heights — one in the rice and

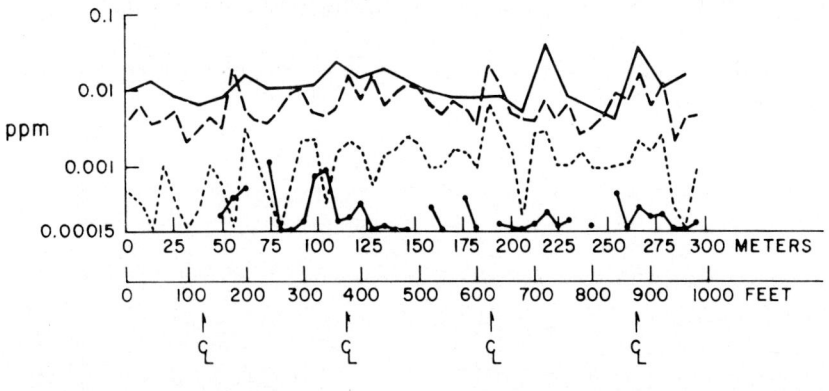

Figure 65. Graph of spray penetration and swath distribution across a rice field, showing ppm (parts per million) of active chemical caught in water-filled cups placed at rice-top level and at water level under the rice, as well as 3- and 24-hour standing water samples (below rice).

one at about the top of the rice. Analysis of water samples for 3 hours and 24 hours after application is shown at the bottom. As can be seen, the rice offered some resistance to spray penetration, but was not dense enough to stop penetration of this large drop size. The chemical levels dropped rapidly, as much as tenfold between the 3-hour and 24-hour samplings, probably due to the moving water. The distribution of the chemical appeared to be satisfactory, and the results obtained on natural mosquito populations were good.

Adult caged mosquitoes and flies are also frequently used in bioanalysis for the effect of spray (Rathburn *et al.*, 1969); however, this is aimed primarily at direct contact with the insects and suited primarily to fine aerosol-type tests. Exposure of fallout sheets that have been placed in the swath of the applicator is still another means for checking the effect of deposited spray on insects.

All types of deposit analysis, whether for quantity, size of particles, or weight of material collected, as well as bioassay on living insects, create somewhat artificial conditions, which can only be rectified by actual checks on natural populations. Thus the importance of any testing programme, short of an operational application, can only be to assist in the establishment of guidelines for the actual job. It is frequently possible, although not always practical, to take deposit and drop size and frequency data during operational runs; but in most cases the best that can be done is to check on the insect, weed, or fungus populations that exist after the chemical application has been made.

11. OPERATIONAL ANALYSIS OF AGRICULTURAL AIRCRAFT USE

The aircraft is basically a tool for use in applying various agricultural materials to aid in food and fibre production. Since it is an expensive tool requiring a considerable outlay in funds, it is essential that every aspect of its use be thoroughly understood and that management of the aircraft with regard to cost and operational data be carefully controlled.

The study of the overall operation, whether it be for pest control, forestry seeding, or for fish-farm spraying, should be carefully done with proper cost-accounting methods, along with integration of control results. This is a systems analysis, or operations analysis, of the specified aircraft mission in relation to the job to be done. Operations analysis tries to find the basic parameters affecting or controlling the productivity of a given system in terms of the job accomplishment, such as total tons of material applied or area covered in a given limit of time.

Tables 3 and 4 list various new aircraft available for application work. But it must also be pointed out that there is a great deal of interdependence between military or paramilitary organizations and various agencies, national and international, which may be called upon on an emergency basis to apply insect-control chemicals or to drop feeds and human foods into disaster-stricken areas. Under such circumstances the cost of aircraft use to perform the job is irrelevant. Also, the types of aircraft used will more than likely be those which are available, and not those most suited economically or performance-wise to the job to be done. On the other hand, commercial enterprises as well as organized government agencies must base their operations on sound management practices and must therefore be highly cognizant of the characteristics and uses of agricultural aircraft.

The choice of aircraft (from those listed in Tables 3 and 4, for example) will then rest with the availability (*a*) of the particular aircraft in a given country as determined by import limitations or local production, (*b*) of good maintenance service for engines, and (*c*) of spare parts, again in relation to import restrictions. While the early aircraft types were basically of military origin, the subsequent changes in engines, airframes, landing gear, and even wing structure and airfoil type point up the need for specific design considerations for application use.

As indicated in Chapter 6, the basic considerations in choosing an aircraft (after availability) are the size of the load to be carried, the distance to landing areas, and the size of area to be treated (Van Bemmel, 1953). Special considerations apply in the case of rotary-wing aircraft, in which case the considerably higher original cost puts an additional burden on the productivity demands, making it difficult to justify the use of more expensive machines without some special features of increased effectiveness of application, either better results or a compensating increase in productivity. As noted in Chapter 6, helicopters can be very competitive in ground-equipment costs, but primarily for those jobs requiring better coverage and crop penetration than can be attained with fixed-wing aircraft. But the helicopter in competition with the fixed-wing aircraft has to do better than this, since specialty jobs such as orchard and vine spraying have not been in sufficient demand or for large enough an area to sustain helicopter use. Hence, the helicopter to be competitive with fixed-wing aircraft must be capable of sufficiently greater productivity to cover the added cost per hour charges needed to pay for the helicopter.

Economics of Operation of Various Aircraft

Table 13 indicates the approximate costs of operating commercial agricultural aircraft in the U.S.A. The fixed costs of maintaining and owning an agricultural aircraft are figured on an annual basis (Norman, 1959). Aircraft operations may lease rather than own aircraft, in which case a lease is usually based in part on the hours of use, as well as on the period of possession.

The depreciation shown under fixed costs in Table 13 is based on an initial cost less salvage value divided over a 10 year period of expected life. If the aircraft is not a total loss by the end of 10 years, it must usually be completely rebuilt. Interest charges on the average value of the aircraft over its lifetime are based on cost less salvage value divided by two, plus salvage value times 10%. The taxes and hangar charges are also based on this average value of the aircraft and are calculated at the given percentages. The fixed costs are based on three different cost classes of aircraft: (*A*) $70 000, (*B*) $35 000, and (*C*) $20 000. These represent a variety of fixed- and rotary-wing types like those in Tables 3 and 4, where approximate cost classes are shown at the far right. No helicopters of practical size and utility are available below $25 000; hence, all helicopters would fall in classes *A* and *B*. The last item of fixed costs is insurance, for which estimated amounts are shown in Table 14. Here, insurance premiums, or rates, are indicated for (1) various ground or storage hazards; (2) collision or crash damage to hull, engine, or other structural elements; (3) liability for damage to persons or to the property of others (not including chemical misapplication or drift); and (4) liability for chemical damage to crops, humans, and animals as a result of aerial drift or direct misapplication. Chemical liability insurance is very costly, being exceeded only by crash coverage, and its cost can be higher or lower depending upon the chemicals that are specifically listed as being covered. Hazardous or highly toxic chemicals, particularly herbicides, are frequently rated for higher premiums, and in some cases the use of a chemical is limited by the availability of insurance coverage. Items 3 and 4, as well as workman's compensation for loss of working time (5), are frequently required of all commercial aircraft operations. As can be seen, insurance becomes a significant part of the costs of agricultural aircraft use.

The variable or operating costs per hour are also shown in Table 13. These indicate fuel and lubrification oil costs, based on fuel only, at 10 to 20 gals (U.S.) per hour, and the costs of repairs and general maintenance, as well as required inspection by approved and licensed mechanics. Major engine overhaul is based on 1 000 hours of use at a cost of $5 000 (*A*) to $3 000 (*C*), depending on engine size. The costs for a major overhaul of radial engines are about 30% greater than for a comparable in-line or flat type and thus raise the hourly cost where radial engines are used, as on the Grumman Ag Cat and the Aero Commander Thrush. The pilot allowance is figured at $15 per hour, but a share in the net proceeds or yearly income is frequently allowed the pilot as well.

Total costs per hour as shown in Table 13 are the sum of the fixed yearly costs divided by the total hours of use plus the hourly operating costs. These are shown for each of the three cost classes. The fixed charges are seen to be highly dependent on the seasonal or yearly hours of use. A low use factor of 150 hours per year shows a fixed cost of only $122.35 per hour for the class A aircraft. But if the hours of use can be raised to about eight times this amount, or 1 200 hrs per year, costs go down in proportion to about $15.29 per hour. Also, the significant difference in the fixed costs in relation to the initial or new aircraft cost should be noted.

When the hourly fixed costs are added to the variable or operating costs, the total hourly cost is found. Since the variable costs are the same per hour, the total cost basically reflects the differences in the fixed costs for each cost class of aircraft as well as the hours of use per year. For example, the difference in total cost per hour between class C and A aircraft is a little over double at 150 hours of use per year, but drops to only a little more than a 50% increase when each is used for 1 200 hrs per year. Thus the essential need for high use is most clearly shown for high-investment aircraft (Janssen, 1959).

Operational Mission

The operational mission and the relationship to several parameters controlling the productivity or area covered per unit of time are as follows: productivity is a function of aircraft load, rate of application, swath width, field or application length, and loading time and ferrying time. In equation form this is expressed as follows:

$$A/T = \frac{Q_L}{(T_R/60)+(D_F/V_F)+(KQ_L/Q_A S_W)(1/V_S+T_T/60D)}$$

A/T = Productivity in ha/hr (acre/hr)
Q_L = Aircraft load in kg (lb)
Q_A = Application rate in kg/ha (lb/acre)
T_R = Loading time (min)

T_T = Turning (end of swath) time (min)
D_F = Ferry distance in km (mi)
D = Field or run length in km (mi)
S_W = Swath (flagged) width in m (ft)
V_F = Ferry speed in km/hr (mi/hr)
V_S = Field speed in km/hr (mi/hr)
K = Constant for metric is 10; for English and U.S. 8.25

This and other similar equations (Janssen, 1959; Amsden, 1962; Van Bemmel, 1953) have been used to describe the aircraft operational mission; however, it would be impractical to substitute values into the equation for each operational situation, so the data are fed into a computer and the output can be either tabulated or graphed, as shown in Figures 66-71 for metric measurement and Figures 72-77 for English and U.S.A. units. Figures 66-68 are for three application rates with fixed-wing aircraft, and Figures 69-71 are for three rates with rotary-wing aircraft or helicopters. Similar data are shown in English units for fixed- and rotary-wing aircraft in Figures 72-74 and Figures 75-77, respectively.

Table 15 shows the operational constants for each of the graphic figures. These median numbers were used as constant values in the equation as each factor was varied separately through the range shown in the graph; that is, when loading time was varied to determine its effect on productivity, then all the other values in the equation were held constant to the numbers shown. The principal difference in the mission of each graph is seen to relate to the application rate, but changes in the other factors were also made in accordance with the application rate, especially for the ultra-low-volume (ULV) type of applications characteristic of the operations shown in Figures 68, 71, 74, and 77.

In each graph the vertical scale on the left shows productivity in hectares per hour (acres/hr), while the horizontal scales are from top to bottom: (1) ferry distance in kilometres (mi); (2) field-run length in kilometres (mi); (3) swath width in metres (ft); (4) aircraft payload in kilograms (lb); (5) application rate in kilograms per hectare (lb/acre); and (6) loading or turnaround time in minutes. Figures 66 and 72 are for fixed-wing aircraft at heavy application rates. Examination of each of the plotted graphs shows the effect on pro-

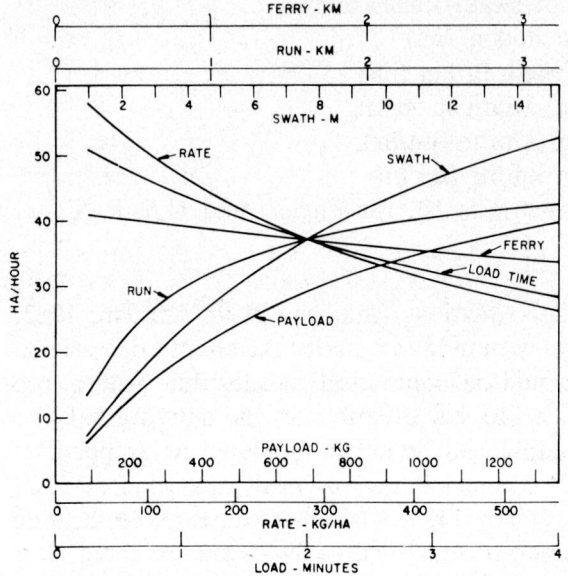

FIGURE 66. Operation analysis (metric) for a fixed-wing aircraft at a heavy application rate.

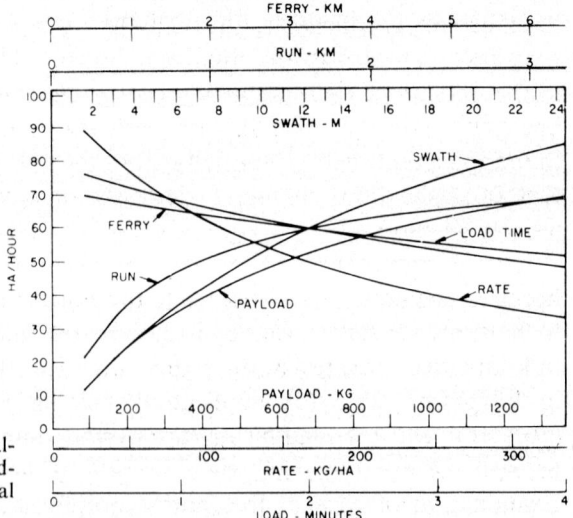

FIGURE 67. Operation analysis (metric) for a fixed-wing aircraft at a normal application rate.

ductivity that changing each factor level produces (all others remaining constant). Thus, if the rate of application (topmost graph) is varied from 100 to 500 kg/ha (89 to 447 lb/acre), the productivity drops from about 50 to 30 ha/hr (45 to 27 acres/hr). The next graph down, for loading or turnaround time, shows a lesser effect on productivity as the loading time is increased. Varying the ferrying distance from 0 to 3 km (from 0 to 1.86 mi) has little effect on productivity, as can be seen in the third graph down. The fourth graph for field run or swath length rises very rapidly as the length is increased from 0.1 to 1 km (.06 to .62 mi), but levels off as the length is increased much beyond this distance. The swath width in the second graph from the bottom shows a rapid increase as the width is increased, rising from about 10 ha/hr at 2 m (24.7 acres/hr at 6.56 ft) to about twice this value at twice the swath width. The last graph is for the effect of payload on productivity and shows a rather steep rise as payload increases from 200 to 600 kg (440 to 1 320 lb), with a lesser rise from 600 to 1 200 kg (1 320 to 2 640 lb). In summary, increasing the application rate reduces productivity sharply, which of course favours the use of concentrated materials or low-volume applications. Increasing the swath width increases productivity sharply, owing primarily to the fineness of the spray-particle size, which gives a wide-distributed swath, and to the length of the aircraft boom. The other factors do not have as great an effect on productivity, but whenever they can be changed favourably, there will of course be a corresponding productivity increase.

Figures 67 and 73 are for a more usual application rate of 50 to 300 kg/ha (45-270 lb/acre), equivalent to a liquid application rate of 50-300 l/ha (5.35 to 32 gals/acre). Within this range of application rate the productivity is shown to change from 80 to 40 ha/hr (200-100 acres/hr). Again, the swath width can be seen to change productivity significantly, while other factors have less effect.

The graphs in Figures 68 and 74 are for ULV application levels and show the strong effect on productivity of a sharp reduction in the application rate. Thus, at lower rates of 0.4 to 2 kg/ha or l/ha (.043 to .214 gals/acre) the productivity (scale A) drops from 1 050 to 725 ha/hr (425-294 acres/hr). The swath width is greatly affected by the fine atomization of ULV applications, and as swath width is varied from 50 to 250 m (160-820 ft), the productivity (scale B) rises from 600 to 2 100 ha/hr (1 482-5 187 acres/hr). These productivity rates

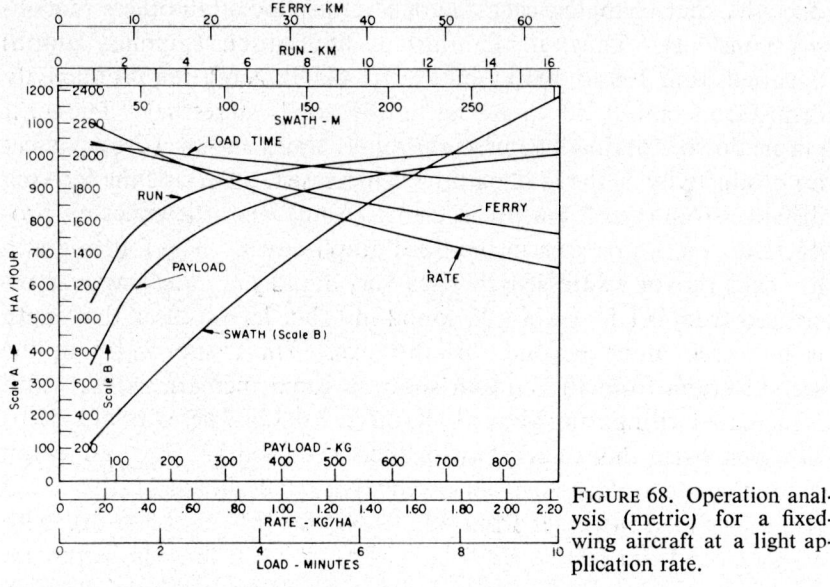

FIGURE 68. Operation analysis (metric) for a fixed-wing aircraft at a light application rate.

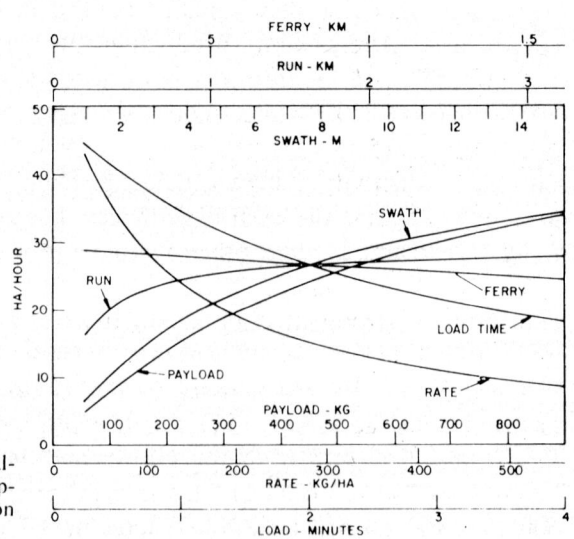

FIGURE 69. Operation analysis (metric) for a helicopter at a heavy application rate.

are characteristic of ULV for large-area application work in control of locusts, grasshoppers and various disease vectors. The productivity reflects the use of wide swaths obtained by wind drift in this type of work.

Figures 69-71 (as well as Figs. 75-77) likewise serve for productivity analysis of helicopter applications. Here, the payload, swath width, and application rate affect the productivity significantly where

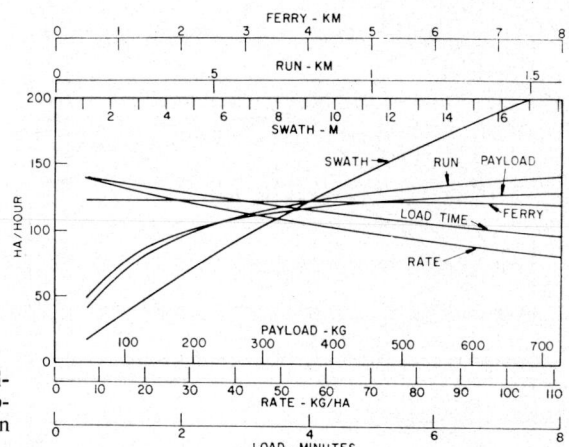

FIGURE 70. Operation analysis (metric) for a helicopter at a normal application rate.

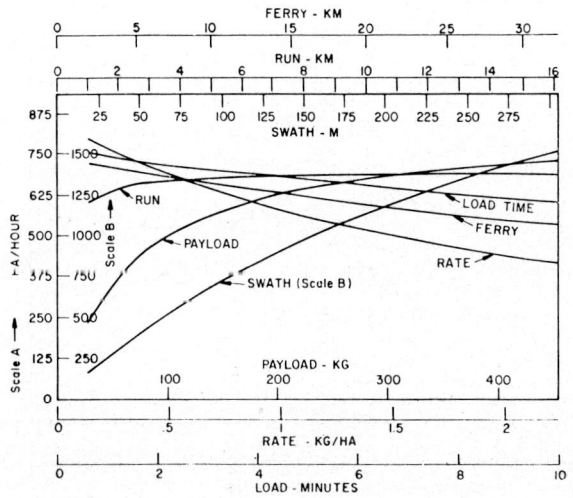

FIGURE 71. Operation analysis (metric) for a helicopter at a light application rate.

high application rates are used. As the application rates are reduced as shown in Figures 70 and 71, the swath width becomes the dominant factor affecting productivity.

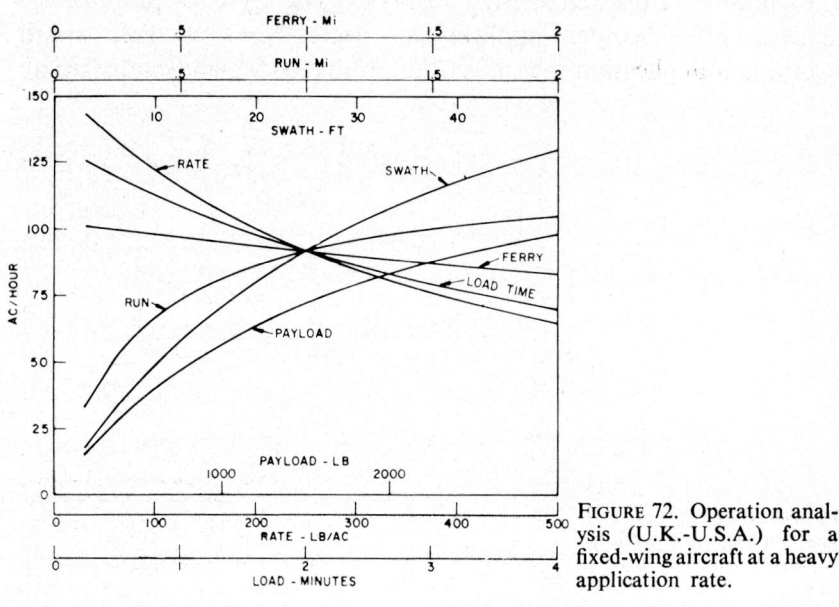

FIGURE 72. Operation analysis (U.K.-U.S.A.) for a fixed-wing aircraft at a heavy application rate.

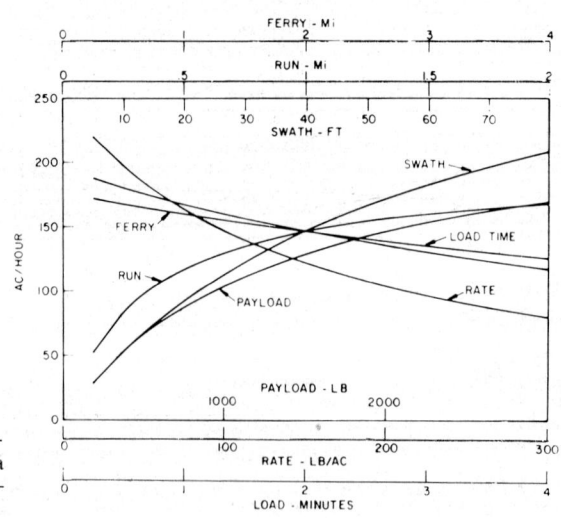

FIGURE 73. Operation analysis (U.K.-U.S.A.) for a fixed-wing aircraft at a normal application rate.

The general productivity of the helicopter compares very favourably with that of the fixed-wing aircraft in the normal operation range of 50-100 l/ha (5.35-10.7 gals/acre) but, as would be expected, drops

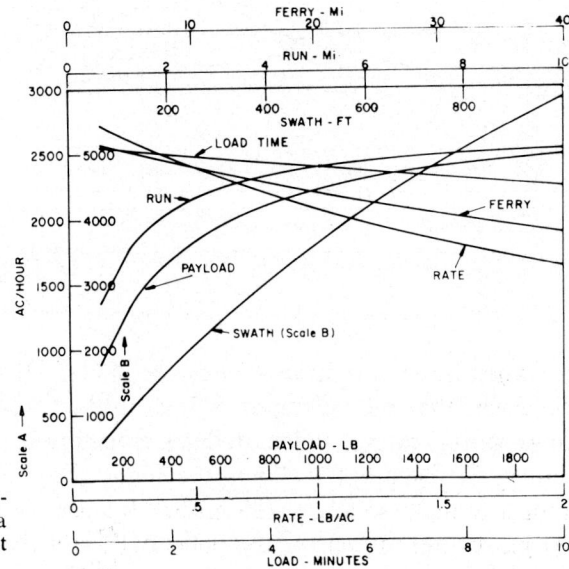

FIGURE 74. Operation analysis (U.K.-U.S.A.) for a fixed-wing aircraft at a light application rate.

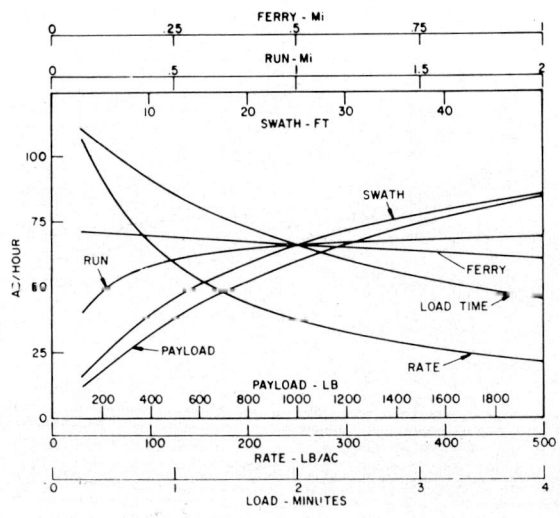

FIGURE 75. Operation analysis (U.K.-U.S.A.) for a helicopter at a heavy application rate.

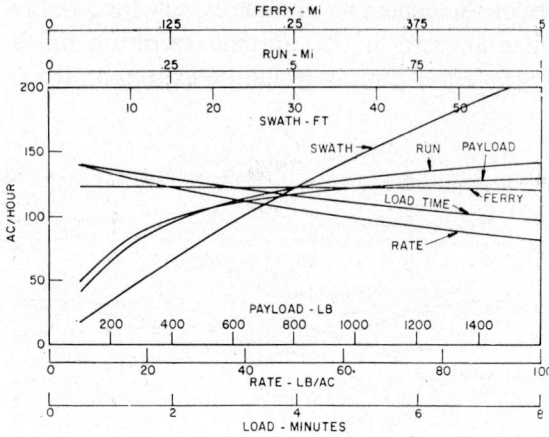

FIGURE 76. Operation analysis (U.K.-U.S.A.) for a helicopter at a normal application rate.

off when heavy application rates are made. It should also be noted, however, that the helicopter is less productive at ultra-low, or ULV, application rates, which reflects the lower operating speed not shown in the graphs. Thus the flight speed as shown in Table 15 indicates that 145 km/hr (90 mi/hr) was used for the fixed-wing versus 97 km/hr (60 mi/hr) for the helicopter, and the difference in productivity between Figures 68 and 71 also reflects this decrease of about 30%.

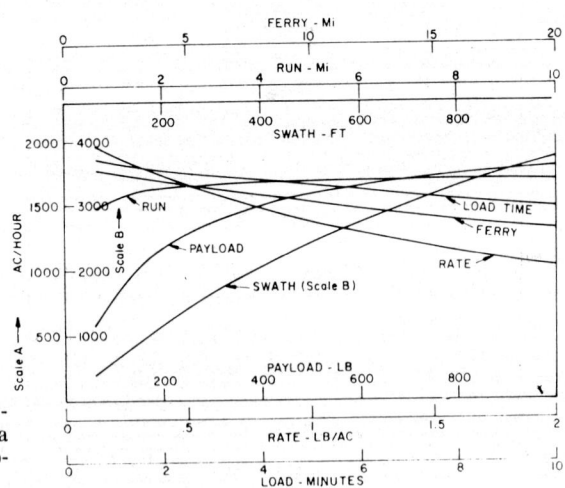

FIGURE 77. Operation analysis (U.K.-U.S.A.) for a helicopter at a light application rate.

Cost of Aircraft Application

From the operational data shown so far the actual per hectare (or per acre) cost of application can be seen to involve three elements: (1) type, size, and initial cost of the aircraft; (2) management for maximum use and minimum cost per unit of area covered; and (3) large-area contract flying versus small farm jobs. The third element relates to the second insofar as management is concerned, but has a significant effect on the cost of large-area applications under contract or by agreement, which greatly affects the basic price per unit of area covered. It is clearly understandable that with an operation calling for high capital investment and high productivity operation the more use hours which can be guaranteed or planned for will make possible a unit cost of operation that can be very favourably low. This guaranteed use frequently involves large-scale applications for vector, locust and grasshopper control, special control measures such as those for rodents and birds, and also such large government-sponsored operations as the top-dressing work in New Zealand (Alexander and Tullett, 1967) and the spray applications on rice in Indonesia (CIBA, *The principles of waterless spraying*, 1969). Still other large-area control programmes involve, for example, applications for control of overwintering insects where there are particularly vulnerable populations in a semidormant or nonreproductive stage. Such a programme is exemplified by the several thousand acres of spray work done each year in California for control of the overwintering grape leaf hopper, a vector of a virus disease of sugar beets.

Other large-scale programmes involve large forest and rangeland applications and are occasionally aimed at the eradication of certain insects invading a new crop area or attempt to eradicate old insect pests. Under these categories large forest spray programmes in the U.S.S.R. and Yugoslavia, as well as in Canada and the U.S.A., have provided an opportunity to control pests by aircraft that would have been virtually impossible with ground equipment both from the economic and the feasibility point of view. The use of biological control materials such as virus and bacteria for insect control, the release of live insect predators and parasites, and large-scale programmes of sterile insect release have also been very well adapted to the use of aircraft.

It is important to note, however, that such large-scale projects as these differ greatly in management practices from agricultural work, which can be highly variable and also difficult to organize in order to obtain maximum cost reduction. For example, when rice seeding is done in a local or nearly uniform climatic area, all the land is ready at the same time and the great pressure of demand forces aircraft operators to have equipment capable of finishing this job within a week to ten days. This means that the excess equipment will not be used as heavily for the remainder of the year, with a consequent low overall use factor and heavier than necessary fixed costs. These costs must then be passed on to the farmers who demand this type of service.

Table 16 indicates some representative charges for various forms of materials applied by aircraft. The three principle cost-controlling factors shown are (1) particle size of the material to be applied, (2) swath width (flagged) to be used, and (3) volume to be applied per hectare (or acre). These data are based on single-engine (maximum 600 hp) fixed-wing aircraft. Smaller ones are usually used for the lower rates of application, while larger, more powerful aircraft are used for heavier application rates, as for top-dressing work. Fine sprays or aerosols at low volume permit very wide swath applications and high productivity in hectares (acres) per hour. This plus the large contract type of operation permits very low charges, as shown in the column at far right. As the spray particle size is increased, as is required for controlled drift applications, the swath width narrows and the application rate must be increased to maintain coverage. Finally, the use of very viscous or thick sprays, such as viscous emulsions and cellulose additives, requires a high application rate, as for brush control, as well as a higher cost per hectare (acre).

Application costs for dry materials are also shown. In general, the dry materials formulated for chemical pesticides are applied for about the same or a slightly higher cost per unit weight. Thus 2 kg/ha is equivalent to a 2 l/ha water-base application, so the charges would be similar. As the rate per hectare increases, the cost goes up; however, for the last rate shown, which is for the application of top-dressing or fertilizers, the cost is much lower than for an equivalent liquid application, partly due to the use of unique loading and handling equipment for large quantities of bulk fertilizer materials at a considerably reduced cost per unit of weight.

Helicopters under competitive conditions may be able to operate at about the same or somewhat higher rates, but in order to do so the productivity must be kept high to finance the extra costs of purchase and maintenance. In some instances, such as orchard and vine work, where helicopters are able to outperform fixed-wing aircraft, cost becomes a secondary concern, and the helicopter is in fact no longer competitive with fixed-wing aircraft, but more competitive with ground-operated equipment.

The operations marked with asterisks are for large-scale contract work, which could involve 1 000-100 000 ha (2 470-247 000 acres) or more. Where a large contract involves a known area and charges (plus low volumes), the aircraft operator can reduce the charge per acre to very low rates as shown. The other operations are based on smaller lot sizes, such as the more usual farm fields, and the charge per unit area is found to be considerably higher because of this and the generally higher volumes used.

These charges relate to competitive aircraft operations in industrialized countries. It is to be expected that in more remote areas, where spare parts are unavailable and maintenance difficult, the costs are also significantly higher.

Calibration of the Aircraft Applicator

It is essential to calibrate or adjust the flow rate and controls on the aircraft to obtain predetermined application rates. When large volumes of either liquid or dry materials are being applied, small errors in amounts used are not likely to be troublesome. However, when the applications become increasingly concentrated or lower total volumes are used, the dangers of overdosage, which may damage crops, or underdosage, which may result in poor control, must be avoided. In order to ensure precise rates of application, it is customary to set the equipment as closely as possible to the desired rate and then to make actual known field length runs either with the material to be used or a suitable simulated material having the same flow characteristics.

The total output or discharge rate of material per minute from the aircraft will be related to the desired application rate per hectare (acre) and the number of hectares (acres) being covered per minute. Thus it is first necessary to establish the basic factors controlling these. The aircraft speed and the swath width (actual or flagged) covered will control the hectares (acres) covered per minute. The aircraft speed will be that recommended for the particular aircraft, engine size, altitude, and load being carried and will range from 80-120 km/hr (50-75 mi/hr) for helicopters up to 200 km/hr (125 mi/hr) for small fixed-wing aircraft. The swath width will vary depending upon the particle size and consequent spread of the material applied. Table 16 shows the range of swaths used, from 305 to 9 m (1 000 to 30 ft). The wider swaths are obtained with aerosols and fine sprays, and the very coarse sprays produce swaths that basically relate to boom length. Figures 54-57 show certain actual spray-swath patterns obtained with fixed- and rotary-wing aircraft. Swath widths for dry materials are a little more unpredictable than for sprays; also, the actual rate applied per hectare (acre) is frequently altered with dry materials by changing the swath width. Figures 43-46 show typical swaths of 10-20 m (30-60 ft) obtained with granules and seeds. The patterns are primarily a function of aircraft size (wing span) and design of the ram-air or spinner-type spreader.

In order to make the final decision on proper swath width, it is frequently desirable to actually fly the material to be applied over a collecting surface, such as a length of wood boards (perpendicular to the flight path) or other surfaces on which heavy oil or grease has been spread to catch dry particles and seeds, or to apply a coloured spray to suitable paper strips or cards laid perpendicular to the flight path. For more sophisticated measurements the collection means shown in Figures 63 (dry) and 64 (liquids) are used.

When the swath width to be used is established and the aircraft speed is known, the hectares or acres per minute can be found by using the the following formula:

$$ha/min = \frac{SV}{600} \qquad \text{Metric} \qquad (a)$$

$$acres/min = \frac{SV}{495} \qquad \text{U.S. or U.K.} \qquad (b)$$

(a) For metric $S = $ m and $V = $ km/hr.
(b) For U.S. or U.K. units $S = $ ft and $V = $ mi/hr.

Once known, the hectares or acres per minute (SV) can be multiplied by the rate of application (Q_A) in kilograms or litres per hectare (pounds or gallons per acre), and the quantity or total discharge per minute (Q_T) by weight in kilograms (or pounds) or by liquid volume in litres (or gallons) per minute can be determined as follows:

$$Q_T = \frac{Q_A SV}{600} \quad \text{Metric} \quad (a)$$

$$Q_T = \frac{Q_A SV}{495} \quad \text{U.S. or U.K.} \quad (b)$$

(a) For metric units Q_A = kg or l/ha.
(b) For U.S. and U.K. units Q_A = lb or gal/acre.

For dry materials the discharge rate thus obtained must be the rate that is actually released from the aircraft in flight. In order to establish this, actual flight with ram-air spreaders must be made, or a rough approximation with spinning spreaders can be made by operating the machine on the ground.

For liquid applications the next step would be the selection of nozzles or flow-control orifices in spinner devices. Tables 17-19 show characteristic flow rates of Spraying Systems Co. hydraulic nozzles at given pressures when operated with water. Slightly higher flow rates (15-20%) can be expected for petroleum and other oils or even for inverted emulsions with oil as the continuous phase. The orifice size for the fan nozzles or the orifice and whirlplate for the cone-type nozzles, along with the pressure used, controls the drop size of the spray. Figures 25-28 give some general information on drop size for hydraulic nozzles, and Figure 29 indicates sizes to be expected from spinning atomizers. When the nozzle type and size have been selected, the flow rate per nozzle can be found from the nozzle tables and the number of the selected nozzles needed ascertained as follows:

Number of nozzles = Q_T / Q_{Noz}

Where Q_{Noz} = flow rate (l/min or gals/min) per nozzle

Various means have been used to simplify the equation shown for the area covered using either dry or liquid materials. Tables, slide rules, and other means can make the job of calibration much easier and aid, for example, in selecting the proper nozzles for different application rates.

Tables 20 (a,b) and 21 (a,b), for the metric and U.S. and U.K. systems, show how data on area and application rates can be collated for easy reference and use. Table 20 (a,b) indicates the area covered in field lengths of 1/4 to 5 kilometres and miles for swath widths of 7.5-65 m and 20-100 ft. When a field of a given size is to be treated, these factors can be established, and after the dosage per unit of area is decided, the total amount of material needed (both active chemical and total mixture), the number of swaths required by the aircraft, and the maximum number of swaths per aircraft load can be found.

Table 21 (a, b) shows the area (hectares or acres) covered per minute when the flight speed (left-hand columns) and the swath width (top lines) are chosen or established. Thus, for example, at 140 km/hr (87 mi/hr) — a common flight speed — with a 15 m (50 ft) flagged or usable swath, 2.9 ha/min (8.8 acres/min) are covered. The data in the tables are all linear — that is, for half the speed or half the swath width the hectares or acres per minute are halved. Similarly, doubling either of these factors will double the area covered per minute. The area covered per minute multiplied by the application rate per hectare (acre) equals the total flow rate in either litres or gallons per minute. Dividing this by the number of nozzles will give the necessary flow rate per nozzle at the application rate desired.

To find the number of hectares or acres that can be covered with a hopper load (ha_L or A_L), divide the hopper load (liquid volume or dry weight) by the rate per hectare.

$$ha_L \text{ or } A_L = Q_L/Q_A$$

Q_L = litres or gallons per load
Q_A = litres per hectare or gallons per acre

Now, by checking Table 20 (a,b) for area covered on each swath, the number of swaths that can be taken by an airplane hopper load can be found, and the load can be adjusted so that the aircraft will not run out of material in the midst of a swath.

The time needed for emptying the aircraft (T_L) is dependent upon the hectare or acres per load (ha_L or A_L) divided by the hectares or acres per minute (ha_T or A_T):

$$T_L = ha_L/ha_T \text{ or } A_L/A_T$$

Also

$$T_L = \frac{Q_L}{Q_A (ha_T \text{ or } A_T)}$$

The hectares or acres per hour, not considering time in turns or for loading, are calculated as follows:

$$ha_{hr} \text{ or } A_{hr} = \frac{60 \; Q_L}{Q_A \; T_L}$$

As a final check on the calibration, to ensure that no mistakes have been made, it is always desirable to make a flight check by putting a known quantity in the aircraft or by noting the given level in the tank with the system filled. Then, a flight of a given time or distance is made, after which the quantity is again measured by draining or by refilling to the original load, and the total amount used for application to the given area will indicate the accuracy of the calibration. Small alterations, up or down, can generally be made by changing pressure, but if a large error is found, the nozzle size may have to be changed.

With the aircraft properly calibrated, the next step is to lay out the procedural flight pattern that will be used to do the job.

12. FLIGHT PLANNING, AIRCRAFT LOADING, AND FIELD LAYOUT

For agricultural operations it is customary for the aircraft operator to become very familiar with the farmer-customer's fields, which he may treat several times a year and even for many consecutive years, perhaps on several crop rotations annually. The operator will usually work from the same landing field, where he may have his aircraft storage, repair, and maintenance services, as well as his headquarters with chemical handling facilities and communications centre, frequently having his own radio equipment and industrial radio frequency assignment.

From such a central location, with all-weather paved runways and frequently nighttime landing-light facilities, the operator will work the surrounding area perhaps to a distance of 6-10 km (4-6 mi). For jobs farther afield he will move the specially designed liquid mixing (Figs. 59, 61) and dry-bulk handling transport vehicles (Figs. 38-40) by highway to remote landing strips, where a handling and loading operation can be set up for another working area of 6-10 km (4-6 mi) in diameter. At the remote landing strip, facilities comparable to those of the home base may be provided, but more frequently with only a dirt or grass landing strip 400-500 m (1 310-1 620 ft) in length and perhaps an irrigation or drainage ditch for water supply. Radio communication will be continued at all times in order to inform the headquarters of the progress of the work, to maintain surveillance of weather conditions and unusual hazards of operation, and to direct the aircraft and loading crew to another landing strip and fields to be treated.

In such a highly organized commercial operation the headquarters will have good detailed maps of the entire area on which applications may be made, as well as local or even specific crop-location maps of the individual farmer's fields and cropping plan. Special note will be made of difficult fields to work, such as those crossed or bounded by power lines, trees, buildings, hills or other obstructions. Whenever a new field is to be treated for the first time, an inspection will be made, first from the ground, to locate any obvious problem situations, and then, if possible, from the air by the pilot who will fly the job, to determine the field shape and its position with respect to prevailing wind, sun glare, and obstructions. When these inspections are completed, the information will be relayed to the headquarters, which will give the pilot (if he was not able to inspect the field personally) final instructions for the job.

Since the pilot and mixing crew will frequently be away from headquarters and the direct supervision of the operations chief, the pilot will usually be given responsibility for the individual jobs to be done and will keep records of information as to time and place, specific amounts of material applied, and weather conditions during application. The pilot will check the ground crew on the amounts of material (active chemical) to be used and the total volume of application. He will check the work order requests issued by the operations chief against the material that has been given to the mixing crew for the job and make sure that the material and rate of use are correct.

Finally, the pilot will have to decide on the manner of "flying" the field to be treated. The basic considerations for any treatment are (*a*) to fly the longest dimension of the field, (*b*) to fly with a side wind (perpendicular to wind direction) and to progress toward the wind, (*c*) to avoid flying into the sun when it is low in the sky, and (*d*) to calculate field length, area covered per pass, and volume per pass in such a way as to ensure a given number of passes per load, without danger of running out of material in the midst of a pass. This can be done either with the basic relations shown on pages 135-139 or with field length and area data shown in Tables 20 and 21.

LOADING THE AIRCRAFT

The proper load for the agricultural aircraft is based of course on the manufacturer's design and recommendations, which give a certificated maximum weight. But government regulations usually permit agricultural weight above this basic certification. The agricultural aircraft operating under special and restricted categories usually requires a specific inspection and, in some cases, a flight check at the

FLIGHT PLANNING, AIRCRAFT LOADING AND FIELD LAYOUT

increased weight before it will be cleared for licensing. In the U.S.A., for example, all aircraft operating above the manufacturer's certificated weight are required by Federal regulation to have this additional clearance whether they be new, converted ex-military, or any other kind of aircraft. The increased agricultural load is based on data provided by the manufacturers to the FAA, which in turn provides a listing of recommended maximum weight increases. It is usual to find increases of 25-35% permissible under such restricted certification.

The aircraft will then be marked for maximum load capabilities under standard atmospheric and other operating conditions. However, the actual field loading — including reductions in load that are necessary with poor landing strips, low-density atmosphere, and other limiting factors — calls for the considered judgement of the pilot.

The loading of the aircraft for the application run is the pilot's responsibility. He must use his training experience and judgement to determine what is a safe load for his aircraft when working from a given airstrip. This, in turn, is also dependent upon the type, length and condition of the given airstrip. The paved runway on level ground near sea level provides the best operating conditions, but the use of grass, dirt, or gravel strips is frequently necessary, and each raises take-off and landing problems unique to the particular surface involved. Grass may slow the take-off (or increase the run needed), while bare clay or dirt strips provide hazards of mud when wet and of dust and dirt when dry, which can clog carburettor air filters and generally create a disagreeable dust cloud around the loading area. Gravel and sand on the strip cause pitting of propellers and damage to wing surfaces, particularly lowered flaps and tail assemblies.

The liquid load can be accurately gauged by markers on the tank that are visible from the pilot's cockpit. The visible marker may simply be a clear see-through section on the end of the tank, properly scaled for the tank volume, or a clear tube attached at the bottom and top of the tank in which the liquid level rises to match the tank level. The visibly marked tank also enables the pilot to check his load when in flight and prevents him from running out of material during a pass across the field. The density of the liquids used will usually be less than that of water, since the petroleum-base solvents and many of the hydrocarbon chemicals are lower in density than a like volume of water (Table 8); however, some solutions, particularly of dissolved solids in water, will have a higher density, and technical concentration materials (used in ULV work) may be heavier than water. It is therefore desirable to check the liquid density and to be careful that an excessive load is not put onto the aircraft accidentally by using the same volume but of a higher-density material.

For dry materials it is customary to use a "skip," or transfer, loader (Fig. 38), which is gauged to one maximum hopper load for the aircraft under given operating conditions. In some cases, hydraulically actuated loaders are provided with an oil-pressure gauge calibrated to the weight of the material in the loader tanks. With such a gauge, a check on the actual weight put onto the aircraft is made for every load. Particular care must be taken with dry materials, to make sure that the aircraft hopper is empty before the new load is put on. With coarse, unprocessed bulk materials, such as those used for top-dressing, a portion of a load may be "hung up" in the hopper without the pilot's knowledge. A quick check before adding the new load can spare the pilot the embarrassing situation of reaching the end of the runway before becoming airborne.

Field Layout and Application

The actual manner of flying a given field is shown in Figure 78. Here, the pilot is shown lining up on two flagmen on opposite sides of the field and dropping into the application swath from his ferrying height for the first pass across the field. The flagmen will try to stay out of the spray pattern by stationing themselves back from the start and stop of each swath. The pilot will not turn on his spray (or dry material) until he is in the application position, and he will turn it off before pulling out for the turnaround. In this way the ends of the field are not covered, but are "finished off" after completing the main body of the field by flying each end (at 90 degrees to the field flight), with one or two passes, depending on the distance back from the ends that the pilot has had to leave. This will in turn depend on the height of obstructions at the ends of the field.

When the pilot approaches the end of a pass, he will (*a*) advance the power setting, (*b*) shut off the material discharge, and (*c*) pull up sharply 15 to 30 m (50 to 100 ft), turn first about 45 degrees away from the next pass, and then bring the aircraft around in the opposite direction to complete the 180-degree turn and line up for the approach to the next swath (Fig. 78). As the aircraft is brought into the applica-

FLIGHT PLANNING, AIRCRAFT LOADING AND FIELD LAYOUT 145

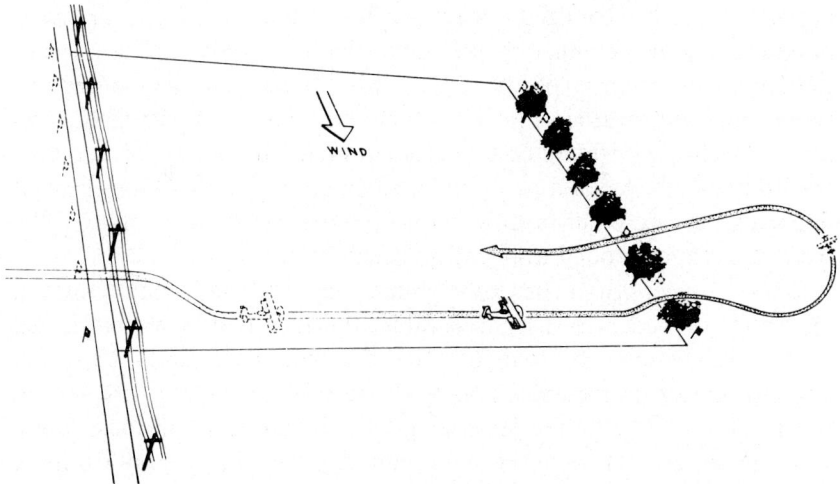

FIGURE 78. Schematic drawing of the manner of flying a field for aerial application.

tion pattern, lining up on the new position of the flagmen, the throttle setting is reduced and, finally, the material discharge is turned on for the application pass. The field may be covered by progressing to and fro until the job is completed; however, it is frequently desirable to divide the field in half and progress as shown in Figure 48 in the "round-robin" or "racetrack" pattern. This second type of pattern is particularly useful where two aircraft are flown, as for applications of top-dressing and other work involving large volumes. One drawback to the round-robin technique is the problem of flagmen. Since the two passes in opposite directions are some distance apart, it may be necessary to station four flagmen instead of only two flagmen, as shown in Figure 48 on page 83.

FLAGGING OR SWATH MARKING

To obtain uniform application coverage of a field or area being treated by aircraft, suitable means must be used to mark the swath patterns (Fig. 48), or "flag" the field for the pilot. The accuracy required depends upon the type of treatment to be made, which relates primarily to the purpose and the physical and particle size of the material being applied. Thus applications of granular materials or very coarsely atomized sprays, such as those used for applying herbicide materials, must be laid down with a highly accurate swath measurement because of the dosage variations that may occur between adjacent swaths (Fig. 47). The variation tolerance frequently limits the choice or use of certain types of materials where, for example, the difference between a dose sufficient to control weeds and one that damages the treated crop may be on the order of plus or minus 20% from the recommended application rate.

While concentrated or low-volume applications require careful metering for accurate discharge rates, the fact that finely atomized sprays are customarily used for ULV applications reduces the necessity of accurate swath widths, as finely atomized sprays will be spread much more widely than coarse sprays by the aircraft and local crosswinds (Figs. 54, 55); note the sharp cutoff of the coarse spray from the jetstream nozzles in Figure 55.

Since it is common practice to use coarse aerosols or finely atomized spray for large-scale forest, tsetse, locust, grasshopper, and a wide variety of vector control applications, the need for accurate swath marking in this kind of work is generally reduced (Armstrong and Randall, 1969). In other instances, where large volumes of top-dressing materials are applied, the accuracy of application is not considered very demanding, since materials will move in the soil, especially on hillsides, and the increased costs of accurate flagging are not warranted in light of the potential benefits to be gained (Alexander and Tullett, 1967).

Several means are used for swath marking, probably the most common being the use of two or more men actually "chaining," or measuring off, the required and predetermined swath distance on each side of the field being treated (Whittam, 1962). As long as the aircraft pilot can see the two men, he can align himself on them and progress easily on the job. The swath or field-length distance is usually limited, however, to about 3 km (2 mi), and longer swath runs require a man in the centre of the pattern. It is customary to use flags of bright-coloured material such as yellow or orange, which will stand out against a green background, and approximately 1 metre square, or 3 by 3 feet. Strong lights such as battery-powered military-type signal lights, which can be directed on the pilot by the flagmen, may also be used for both day and night flying.

Fixed flags are frequently set up in fields that are repeatedly covered over several years, such as orchards, vineyards, bananas, and pineapple, as well as certain specific blocks of insect-infested forests, swamps, or waterways. In order to vary the swath centre and ensure successive application uniformity — in orchards, for example — the flag may be aligned directly under the aircraft on one mission and aligned on the wing or rotor tip on another mission.

The problems of flagging become more complicated for operations conducted in hilly or mountainous areas. Here, the application flight technique may be either to fly on the hill contour, so as to maintain relatively uniform elevation, or to fly up and down the hills, as is normal practice on flat land. In the latter case, the flagger will have to stay on the crest of the hill in order that the pilot may see him. When highly toxic materials are being applied, this can be a serious hazard to the flagman, who will try to move quickly to the upwind, or new flag position, as the aircraft approaches him, but cannot always avoid being at least partially exposed. This is most likely to happen under the especially hazardous condition of no wind and temperature inversion.

Besides fixed flags, various temporary and semifixed markers have been used, such as balloons, paint, coloured dust, smoke bombs, and aerosols produced by engine exhaust, flat cards, boards, or coloured cloth (poor low-level visibility) and the rather unique toilet-tissue dispenser, which has found considerable favour for applications in hilly areas (Whittam, 1962), as the tissue paper leaves a conspicuous marker on the tops of plants or trees. But any markers of this sort, including coloured dyes and dusts dispersed from the aircraft, are of greatest value on long passes over hilly or mountainous land and still require some type of initial swath measurement, such as flags or flagmen. For forest area work balloons have been widely used, either the round type or the more aerodynamic shapes (Kytoon) which reduce wind drag and show less tendency to be dragged down in a high wind; however, wind socks and flags mounted on long poles or in tall marker trees are generally found to be more useful than balloons (Whittam, 1962).

Navigational instruments would at first glance appear to offer a possible solution to the problem of swath-width control, but the frequency of turns and the relatively short swath lengths customarily found in agricultural work do not allow the pilot time to respond to such information as compass bearings or other common navigational aids. For large-area forest, vector-control, or other type applications, where swath length may be as long as 40-60 km (24-37 mi), an initial swath measure by flag in addition to a compass bearing for the aircraft has been used with modest success, primarily to reduce the need for intermediate flaggers in the long swath.

Probably the most sophisticated means for swath measurement and maintenance of proper course is the use of electronic navigational equipment, such as the Lorac, Loran, Decca, and Raydist types. Many air and sea navigation systems of this sort are available, but the most widely used thus far is the Decca system, probably due to its rather wide availability and a simplified unit called Hi-fix or Sea-fix. This unit is composed of two mobile field stations which provide an electronic tracking signal that activates the aircraft-mounted tracking system (computer and flight log) and a turn "left-right" guidance meter for the pilot to follow (Joyce *et al.*, 1968). The stations are set from 25-50 km (16-31 mi) apart and perpendicular to the proposed flight path. The aircraft is guided on the "lanes" of prescribed swath width, with the pilot following the "left-right" turn meter. Accuracy of swath width to 2-3 m (6-10 ft) is reported, constituting a less than 5% error on an 80 m (262 ft) swath. This is a significant improvement over land-based flagmen using balloons, lights, or marker flags; however, for agricultural work involving the use of controllable placement sprays, rather than coarse aerosols, the error of 3 m in 10 m would not be acceptable. Thus the present electronic navigational aids have the greatest potential for large-area applications where finely atomized sprays in wide swaths can be used or where swath accuracy is not essential. Continued improvement and reduced lower costs are foreseen as electronic navigational equipment is developed. In the meantime the use of flagmen on the ground will continue to be the basic technique for swath marking.

Field Mixing and Handling of Hazardous Materials

The handling, mixing, and loading of materials at the airport or remote airstrip constitute an important part of the operational mission. As noted above, the pilot is generally made responsible for checking the field mixing and handling of chemicals, but the actual work will be done by a crew of two or more people who will follow the mixing procedures prescribed by the chemical manufacturer, load the airplane to its proper capacity, and then proceed to mix another batch while the aircraft is making its application run.

Since the time for loading subtracts from the field efficiency of the operation, it is desirable to keep this time to a minimum that is consistent with the overall operation. Rapid mixing and handling facilities are highly desirable, and good mixing equipment is as important as the aircraft application equipment.

Dry materials — granules, seeds, and large particle baits — are most efficiently handled by bulk methods and are not mixed in the field. Any blending or mixing, such as adding herbicide granulars to fertilizers, is done in central mixing areas and hauled in large-bulk boxes or in transport vehicles as shown in Figure 38. Bulk top-dressing materials may be hauled to the loading site in transport vehicles and dumped in piles to be picked up by the "skip" loader device as shown in Figure 39. Still another system for bulk fertilizers, baits, and other dry materials is shown in Figure 40, where large bags (one aircraft load) are loaded at a central area and transported to the loading site on the vehicle illustrated.

Seed and fertilizers (except when chemically treated) are generally non-toxic to personnel, but certain chemicals may be quite toxic, even in granular form, in which case both the loading personnel and the pilot should wear protective goggles and respirators (Bailey and Swift, 1968). In the case of dusts the material is premixed and sent to the field in approximately 25 kg (50 lb) bags. There the bags are opened and put directly into the aircraft or are placed in a skip loader (preferable) and transferred to the aircraft. Again, toxic chemicals require special handling, as will be explained in detail later in this section.

Spray material, with the exception of undiluted technical applications, requires field mixing and handling with special equipment designed for this job. Spray materials may be supplied as emulsions, wettable powders, or in true solution form. The form to be used depends in part on the solubility characteristics of the active chemical. Most chemicals are soluble in petroleum solvents, a few require glycols or alcohols, and a few others are soluble in water. By far the most widely used formulation is the emulsion type — the concentrate forming a normal oil in water emulsion, where water is used as an extender or diluent for the formulation. Wettable powders are virtually insoluble in available solvents, so these are ground up in a dust formulation of 75-90% less than 20 μm diameter, a wetting agent is added along with water, and the formulation is then used as a suspension in water phase. True solutions in water are simply dissolved in the amount of water to be used, while applications using petroleum oils or alcohols as the solvent may be applied without any water diluent, usually for ULV (ultra-low and very low volume per unit area) applications. The emulsion or other concentrate is supplied by the manufacturer in 20 litre (approx. 5 gal) containers or as wettable powders in dry bags, with active or technical ingredient (AI) expressed as a percentage by weight or as grams of active ingredient per litre (1b/gal). The remainder of the material will be the chemical solvent and diluent, the needed emulsifying or wetting agent, and other additive materials, surfactants, or adjuvants that may be included to make the finished spray fluid adhere and spread on plant surfaces. The dosage of active chemical to be applied is usually expressed as grams (1b) of active ingredient (AI) per hectare (acre). Therefore, the mixing instructions to the crew will indicate (a) AI in grams per hectare (1b/acre), (b) total volume of finished spray in litres per hectare (gals/acre), and (c) total litres (gals) of one airplane load for covering a predetermined number of swaths. For example, a 450 g/l AI material to be applied at 900 g/ha in a total of 50 l/ha would consist of 2 litres of emulsifiable concentrate material to every 48 litres of water for each hectare. If the aircraft load for a given number of complete swaths is 1 000 litres, then 1 000/50, or 20 ha, can be sprayed with one aircraft load, and 40 litres of emulsifiable (or other) concentrate will be placed in the mixing tank, which will then be filled to 1 000 litres by adding 960 litres of water. In U.S. units this would correspond to approximately (a) a 4 lb/gal AI material applied at 1 lb or 0.25 gal of concentrate per acre: (b) a 5 gal/acre application rate; and (c) a 250 gal aircraft load, or 50 acres covered per aircraft load. Thus the crew would mix 50×0.25 or 12.5 gals of emulsifiable (or other) concentrate, containing 50×1 or 50 lb of AI, and add 237.5 gals of water to make one aircraft load of 250 gals of material. In practice this would probably be increased to a total of 275 gals, adding the proportional amount of emulsifiable concentrate (EC), so as to have an extra amount of material available in the aircraft above that which is needed for the 50 acres to be covered. The successive loads placed in the aircraft would be accurately measured, the application rate checked, and the tank refilled to the amount calculated.

FLIGHT PLANNING, AIRCRAFT LOADING AND FIELD LAYOUT   151

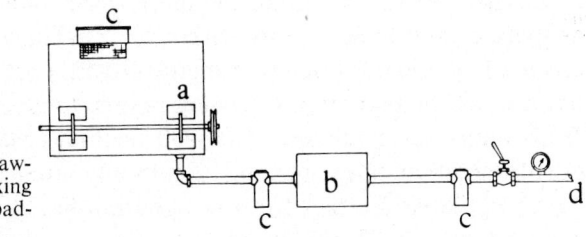

FIGURE 79. Schematic drawing of a liquid-spray mixing tank with mechanical paddle agitation system.

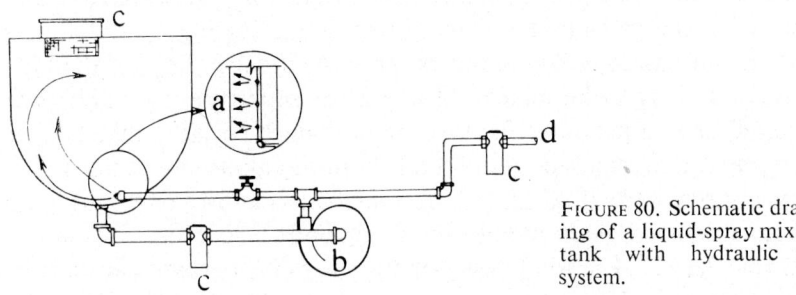

FIGURE 80. Schematic drawing of a liquid-spray mixing tank with hydraulic jet system.

Many applicators have liquid meters mounted on the mixing tank discharge which accurately measure the load placed in the aircraft. For low-volume application, particularly where several spot jobs are done on a single mission, a liquid meter in the aircraft discharge (N in Fig. 6, page 39), placed so the pilot can read it, indicates not only the amounts used on each spot application, but also whether the aircraft is out of material.

DESIGN OF FIELD MIXING TANKS

The field mixing tank for all liquid formulations is shown schematically in Figures 79 and 80. Figure 79 shows a mixing tank equipped with a mechanical paddle agitator system (*a*) driven by a small gasoline engine (not shown), which also drives the transfer pump (*b*). The pump is usually a small centrifugal type capable of 80-130 l/min (21-34 gals/min) at 1.75 kg/cm^2 (25 lbf/in^2). The paddle agitator is usually driven at around 100 rpm for a 30 cm (12 in) diameter blade. Screens for keeping the liquid clean are placed at the tank opening (coarse mesh), at the input to the pump (also coarse), and at the discharge from the pump. At the last point the mesh size should be finer to prevent plugging of nozzle tips on the aircraft. This may vary from a coarse screen of 4 openings per centimetre (10-12 openings per inch), like that used at the two other screening points, to a very fine screen of 40-80 openings per centimetre (100-200 openings per inch), as required for the small nozzle openings commonly used for ULV spraying. Figure 80 shows another form of agitation-liquid recirculation, which may also be used for most all mixing tanks, except when thickened sprays such as methyl cellulose, inverted emulsions, and the like are to be handled. For these, mechanical paddles are more effective and positive in energy transfer. The system design for Figure 80 requires that the minimum energy input be prescribed depending on the type of formulations to be mixed. Table 22(a) shows the jet nozzle size required at two pressures for three tank diameters. Note that the jets are spaced uniformly along the tank bottom (round or rounded bottoms are required) at 30 cm (12 in) between jets and 15 cm (6 in) from the tank ends. The flow rates for the various nozzle sizes are shown in Table 22(b). The total flow rate for the tank size used can thus be found by multiplying the flow per nozzle by the numbers of nozzles used. From the equations on page 46 the size of the pump and driving engine can be found. The flow rates shown in Table 22(b) are for emulsions, solutions and suspensions up to 120 g/l (1 lb/gal) of wettable powders. Higher wettable-powder rates require greater flow, and for thickened sprays the mechanical system is preferred.

Hazards of pesticide chemicals

The hazards of pesticide chemicals to the health of pilot and handling crew vary from the mildly annoying to the highly dangerous. Table 24 lists a wide variety of chemicals used for weed, insect, pathogen, and other pest control. Four categories of health hazard are listed — from the highly toxic organophosphate materials to the least dangerous, including in the last category the widely used DDT. Toxicity of the chemicals is usually established on test animals, such as rats, guinea pigs, and rabbits. Oral, inhalation, and dermal toxicity are all determined. Also, each chemical usually has one mode of entry which has a more toxic effect than the others. The toxicity data taken from test animals cannot however be extrapolated directly to humans, as experiences with human exposure along with animal toxicological data must be used to properly judge the hazards of toxic chemicals to humans (Bailey and Swift, 1968).

Protective clothing, rubber (not leather or cloth) gloves, coveralls and hats of water-resistant materials, and rubber shoes or boots are desirable for mixing and chemical-handling crews when working with any agricultural chemical material, even though such clothing can be disagreeably hot during warm weather and restrict the movements of the wearer. When working with the hazardous chemicals listed in the two left columns of Table 24 extreme care should always be maintained, not only for the loader crew, but also for the airplane pilot and the field flagmen. The pilot should not assist with loading operations, and the mixing and loading crews should wear pro-

FIGURE 81. Worker wearing rubber gloves and respirator placing a toxic chemical in the sump. A jet of clean water will be used (note the pipe at the bottom of the sump) to rinse out the can before disposal, and the rinse goes into the spray tank. In actual performance the worker would be dressed in protective clothing, *not* shirt sleeves.

tective clothing, goggles, helmets, and approved respirators or be equipped with forced-air helmets, now also available with cool-air air-conditioned attachments (Bailey and Swift, 1968).

The chemical concentrate is many times more hazardous to handle than the diluted spray material. Handling systems which provide for pumping from concentrate tanks or barrels and eliminate the tipping or lifting of cans are highly recommended (Fig. 59). Elaborate fixed-base loading and mixing operations may make possible the transfer of correct amounts of concentrate from storage barrels to mixing tanks and thence to the aircraft without the crewman handling any part of the system except the loading hose. Using quick-coupling tank bottom or wing boom loading hose further reduces the hazard of spilling material on the aircraft, pilot, or loader (Fig. 6). Rinsing the emptied container with clean water and pouring the effluent into the sprayer tank is a recommended procedure for disposal of pesticide cans. The rinsed cans can be moved into landfill or recycled for steel mills (Fig. 81).

Emergency procedures should be established for any personnel that may be exposed to highly toxic chemicals. Physicians who may be called to treat workmen should be notified in advance of the type of chemical that is being used, and communication by radio or telephone should be available when handling hazardous chemicals.

Base levels of crewmen cholinesterase (as an indicator of phosphate or carbamate poisoning) taken by the local physician are desirable before the pilot and crewmen work with these materials, and regular checks for cholinesterase levels during the working season or if an accidental exposure occurs ensure that personnel will be protected from receiving toxic doses of these cumulative reaction chemicals.

Cleaning aircraft equipment

Equipment, both aircraft and mixing units, used for all chemicals should be cleaned and maintained on an established schedule. Thorough cleaning with soaps and detergent compounds should be done after each day's (or night's) work, and the effluent from the cleaning process must be drained into a sump or a safe disposal area where no hazard to humans, animals or wildlife can possibly occur. Proper rinsing and disposal of empty pesticide containers must be arranged, usually by rinsing into the mix tank transporting these to approved disposal areas. Containers should not be left in the field, dumped into streams or lakes, or even disposed of by burning as toxic fumes may be released at elevated combustion temperatures.

Certain chemicals — for example, the phenoxy herbicides (2,4-D, MCPA, and others) — are difficult to clean from aircraft and mixing equipment. Traces of these have been known to damage sensitive crops, as when equipment contaminated with 2,4-D was used to spray the crop with another spray chemical. Special cleaning and detoxification are required or, better still, equipment used for 2,4-D should not be used for other applications to sensitive crops. Many aircraft-application operators use steam cleaning with strong washing soda (strong basic as opposed to acid reaction materials) or other neutralizing chemicals. Check-lists for equipment maintenance with regular inspection of equipment for leaks or malfunctioning parts should be instituted for any pesticide application operation. These should be coordinated with the usual aircraft maintenance and inspection generally required by government authorities for aircraft airworthiness and licensing. While the normal aircraft engine and structural inspection are adequate for normal service, recognition and consideration must be given to the more rigorous service and added exposure to corrosion and abrasion experienced by agricultural aircraft.

13. AIRCRAFT FLIGHT SAFETY AND AIRWORTHINESS

By far the greatest concern in agricultural aviation is the safety of the pilot and the aircraft during application work, including take-off and landings (Spuybroek, 1969). The accident rate for agricultural aviation in many countries is found to be highest of all industrial or commercial flying, only exceeded by that for private pleasure aircraft. For example, in the U.S.A. the general aviation accident rate in 1967 (excluding airlines) was 0.27 per 1 000 hrs of flying, for agricultural aviation 0.36/1 000 hrs, and for private pleasure flying 0.39/1 000 hrs. Fatalities occurred in about 9-11% of all agricultural and pleasure aircraft accidents (Steel, 1969). While all aviation accident rates have decreased considerably since the end of World War II, the agricultural rate, while decreasing, has maintained about the same relative position with respect to other flying. Agricultural aviation accident-rate trends in the following table are based on U.S.A., New Zealand, and Australian data and shown in frequency per 1 000 hours of flying.

	1949	*1951*	*1958*	*1963*	*1965*	*1966*	*1967*	*1968*
U.S.A.	—	0.517	—	0.36	—	0.31	0.35	0.286
New Zealand	3.27	—	0.59	0.36	0.31	—	—	0.5
Australia	—	1.5	1.35	0.5	0.4	0.34	0.38	0.35

Reports from other countries indicate that similar accident rates exist and that high rates frequently occurred at the beginning of agricultural aircraft expansion, although fewer statistics are available than for the three countries listed. It is encouraging to see that a steady drop (with regressions in 1967 and 1968) in agricultural accident rates has occurred as aircraft use has expanded.

The causes of accidents are analysed in various ways. Pilot error covers a wide range of responsibility, but it is given as the cause of accidents in 70-80% of all data reported from U.S.A., New Zealand, Australian, and U.K. sources. Errors of other personnel, such as loaders, mechanics, and flagmen, were stated as causing 10-11% of the accidents. Adding this to the percentage attributed to pilot error correlates accident responsibility to a very high degree with human error and careless or poorly managed operations. Power failure was indicated in 16-21% of accidents and structural failure in around 10%.

The pilot error data were further broken down to the following: (a) take-off, taxiing, and landing at 18-36%, correlating to poor airstrips and heavily loaded aircraft with frequent landing and take-offs, as is the case in top-dressing work; (b) collision with ground, trees, wires and cables, and other objects, ranging from 12% to as high as 44% and appearing to correlate to flying in heavily populated areas (highest rate) versus open farmland operations; (c) accidents in connexion with the application run (stalls, maintenance of flying speed, and loss of control) at 14-16%, remarkably the same for the four countries noted. Another survey (Reich and Berner, 1968) broke the data into a few further categories, listing take-off and landing at about 22%, turns at swath ends at 18.5%, swath runs at 15%, pull-up at swath ends at 8.6%, and approaching the swath at 5.6%. Given the very high number of accidents associated with pilot error, it is evident that training and experience under expert agricultural pilots are essential to the beginning and continuance of a career in agricultural flying.

ATMOSPHERIC CONDITIONS

Problems related to air density can also be a source of difficulty for the unwary pilot. Reduced air density affects the aircraft performance in two related ways: (1) it reduces maximum engine power because less air mass per unit of volume is drawn into the carburettor; (2) it simultaneously provides less lift for a given wing configuration, hence reducing the ability to become airborne with a given load or reducing the rate of climb once the aircraft is airborne. Loss of air density occurs with (a) an increase in air temperature, as a higher temperature means an expansion in the volume of a given air mass; (b) increased altitude, due to reduced pressure on a given volume of air; and (c) increased humidity, as air with a high moisture content is less dense than dry air. Most pilots flying a heavily loaded aircraft quickly learn to appreciate the effects of these density factors. Still, a rapid increase in temperature, as from early morning to midday operations, or rapid changes in humidity occurring with a weather front or storm condition can cause trouble for the unwary pilot.

Wind velocity and direction are highly important, not only to the operational flight pattern, but also in deciding whether certain types of applications should be performed or not. Where a distinct hazard is apparent, such as a susceptible crop in the downwind direction from a herbicide application, it may be judicious, if not essential for the safety of the crop and peace of mind of the pilot's insurance company, to postpone the job until the wind direction turns away from the susceptible crop. An increased wind velocity in itself does not increase the amount of downwind transport or drift, as the amounts lost to drift depend on particle size and, in part, on evaporation. But wind does displace the swath pattern in proportion to the increase in velocity. Hence, when working in a crosswind, the displacement of pattern is nearly the same for each pass, so the overall effect is a reasonably uniform pattern. If, however, the wind changes velocity or direction during the operation, the swath displacement will be uneven, so an uneven pattern results (Brazelton, 1970). Increased turbulence with higher wind speeds — especially close to the ground, where most agricultural applications are made — causes increasing difficulty for the pilot in maintaining constant altitude and a fixed direction. Flying crabwise in a crosswind reaches a practical maximum when the wind velocity approaches 25 km/hr (16 mi/hr) and becomes difficult and very tiring at 40 km/hr (25 mi/hr). Making low-level turns or taking off and landing other than into the wind when it exceeds 30 km/hr (18 mi/hr) becomes both hazardous and tiring for the pilot flying a heavily loaded agricultural aircraft.

Wind velocity is not constant with height, but follows a gradient or increase of velocity from ground level (zero) to a maximum at 10-15 m (30 to 50 ft). The agricultural pilot making his turns may accelerate from a level of 10-13 km/hr (6-8 mi/hr) to 25-40 km/hr (16-25 mi/hr) with a change of altitude of 2-10 m (7-33 ft).

GROUND EFFECT IN TAKE-OFF AND LANDING

The take-off and landing operation accounts for the greatest number of agricultural aircraft accidents (estimated as high as 60%), primarily due to heavily loaded aircraft and poor airstrips. But the take-off can also be an especially hazardous operation, as noted above, because of subtle changes in such things as air density and airstrip condition, as well as because of another confusing phenomenon referred to as "ground effect." When an aircraft is flown closer than a height of about one wing span above the ground, the ground reflection of the airstream or downwash from the wing results in a rapidly lowered induced wing-drag as the aircraft approaches the ground, and, therefore, a lowered power requirement to maintain flight and altitude. On take-off the reverse is true: it is possible for the aircraft to become airborne but to be unable to climb out of the ground effect, above 2-5 m (about 7-16 ft) for the usual agricultural aircraft, before running out of runway or a clearing beyond the runway. The effect is noted on landing when a floating of the aircraft can be felt, particularly during the flare-out. The ground effect, then, is a highly desirable phenomenon for both take-offs and landings because of the added buoyancy given the aircraft during this critical stage of flight; but it can be a serious hazard if the pilot of an overloaded airplane finds he cannot climb out of the ground effect on take-off.

STALLS IN TURNS

Accidents related to procedural turns (pull-out, turnaround, and reentry into the swath pattern) rank high on the frequency list, even being rated by some observers as causing more accidents than take-off and landing. As with take-off and landing accidents, there are several factors which complicate the hazards to be encountered in the procedural turns. For example, aircraft load and loss of lift due to reduced air density contribute to changing the stall characteristics of the aircraft which underlie the turn hazard. Agricultural pilots are inclined to treat turns as an undesirable loss of application time; as was noted under operational analysis (see page 124), the short field requiring many turns does reduce the productivity of the operation. Actually, the turn should be considered an essential part of the operation, and the minimum turn, while possibly saving a few seconds, can also magnify a potentially dangerous operational situation.

Basically, the minimum turn radius involves a balance of the total lift producible by a given aircraft wing and the load imposed on that wing by the normal weight of the aircraft, plus the added weight due to the centrifugal force produced by the turn. Increasing the angle of attack of the wing increases the coefficient of lift and thus the total lift of the wing. Increasing the air velocity or aircraft speed also increases total lift. For a given wing design the angle of attack can, however, only be increased to a given maximum before air circulation stops and the wing stalls. At that point, lift drops, controls go soft, and the aircraft may buffet or shake and lose altitude. If at the same time the aircraft is entering a turn, the situation is further compounded by the changing angle of attack of the rising versus the lowering wing, which may cause the lowering wing (higher angle of attack) to stall and results in a spiral dive towards the lowering wing.

The rapidity with which the stall speed is altered by the degree of turn is illustrated by the change in stall speed as turn angle is increased for a given typical agricultural aircraft. Whereas the stall speed at zero bank, or level flight, is 110 km/hr (69 mi/hr), it rises at a 30 degree bank to 118 km/hr (73 mi/hr). This explains very graphically the fallacy involved in reducing speed to tighten a turn. Recovery from a stall normally can be achieved by applying power and levelling the aircraft, but a stall in a turn which has started a spiral dive requires reducing power, levelling of the wings, and keeping the nose down. All too frequently a stall in a turn at the low flight levels practised by agricultural pilots does not permit this manoeuvre, so a crash results.

The apprentice or learner pilot needs many hours of practice to gain the "feel for his aircraft" which becomes a significant part of his ability to handle it in the confines of agricultural work. Needless to say, acrobatics — hammerhead turns, wingover turns, and the like — have no place in agricultural flying.

The helicopter presents a different situation and turns are customarily made by using the forward motion at the end of the swath to gain altitude. This can be adjusted to practically stop the helicopter, at which point the fuselage is rotated 180 degrees, while shifting to the new swath position and then dropping into the new swath pattern. Thus the helicopter can gain time over the fixed-wing aircraft in turns, and also by landing close to the fields being treated; however, application speed is frequently 25-30% less than that of a fixed-wing aircraft of similar size, which tends to equalize productivity between the two.

A very specific hazard for agricultural aircraft is the multitude of communication and power lines that crisscross most of the highly developed crop areas, particularly around population centres. Even though pilots customarily locate all interfering wire before working a given field, it is very easy to lose track of these as they are frequently difficult to see against a dark field. Suggestions for marking these wires include hanging paper streamers or tapes on the lines or, better still, placing large plastic baskets at given intervals under the lines. The problem has also been alleviated in a rather drastic manner by deliberately placing sharpened leading edges on protruding parts of the aircraft, where wires might catch the landing gear, the pilot's canopy, or the vertical stabilizer. The justification for wire cutting is that, if interception of a wire is inevitable, it is better that it be cut than to have it drag the aircraft into a crash.

Pilot fatigue

Pilot fatigue has long been considered a primary factor associated with accident rates. In many countries limitations on the number of hours of agricultural flying per day and per month are prescribed by law in order to protect pilots from the tendency to overextend their physical resources. Contributing factors (Baruch, 1970) to fatigue for the agricultural pilot may be (*a*) frequent early morning operations, thus reducing the pilot's rest; (*b*) low-altitude flying, many obstructions, and frequent turbulent weather; (*c*) long working hours, multiple landings and take-offs (up to 10 or 15 per hour), and poor landing strips, which are frequently dusty and rough; (*d*) high-temperature working conditions and slip stream buffeting in open cockpit planes; and (*e*) exposure to toxic chemicals, which may impair the pilot's vision, increase his drowsiness, or interfere with his sense of balance and direction. Any improvements in aircraft equipment and airstrips to protect the pilot and reduce the physical demands on him will aid in reducing fatigue and pilot-induced accidents. Even with the best available equipment and flying conditions, however, a survey of pilots engaged in agricultural aviation in Israel (Baruch, 1970) noted that the following medical conditions were observed in a group of twenty-four pilots during the 1968 application season: (*a*) generally the group lost weight; (*b*) body temperatures rose as high as 38.5°C (101.3°F) during flights; (*c*) blood sugar and eosinophil cells tended to decrease during the work period, indicating fatigue; (*d*) pilots lost sleep due to irregular hours; and (*e*) pulse-rate electrocardiogram data and blood pressure appeared not to be affected. There was no direct exposure to pesticide chemicals during the season as the pilots did not handle chemicals, but some level of exposure, if only to vaporized materials, must invariably occur when the pilot flies the aircraft and releases his application load. Cholinesterase data taken on the pilot and crew three times during the season did not indicate any significant trends, and since no comment was made by the authors, it can be assumed that no organophosphate poisoning could be demonstrated; however, stress factors relating to the drop in eosinophil, loss of weight, and rise in body temperature were obviously present and could be at least conducive to increased accident rates. There appears to be no question that agricultural flying is an extremely tiring and hazardous occupation. It is thus essential that every consideration be given to reducing this hazard by providing the best possible equipment, the least tiring working conditions, and reasonable limits on duration, with the greatest emphasis possible on accident prevention and reduction. Only through an intensified programme adequately backed by reasonable regulation can the occupational hazards of agricultural aviation be brought to a reasonable level.

STRUCTURAL DESIGN OF AGRICULTURAL AIRCRAFT

The design of agricultural aircraft, while basically the same as for a good utility airplane, must give high priority to rugged construction and structural safety. The accident rate attributable to structural failure is about 10%, which is low considering high aircraft loading and the rough airstrips frequently used. In addition to structural integrity, the agricultural aircraft must provide protection from the chemical load during flight and, insofar as possible, in case of a crash.

Crash protection involves the following factors which are built into the aircraft specifically designed for agricultural use (see pages 28-29 and Baruch, 1970):

1. The greatest possible amount of energy-absorbing structure should be placed ahead of the pilot; the chemical tank, pumps, batteries, and other heavy parts should also be located in front of the pilot.
2. The cockpit should be designed as a major structural element to protect the pilot; the framework should bend outward rather than inward in case of a crash; and both rollover and forward crash protection should be provided in the design.

3. Fuel tanks should be located as far as possible from cockpit and engine — in the wings, for example.
4. Seat and shoulder belts with inertia-type reels and crash helmets should be mandatory.
5. Closed, dustproof cockpits provide further pilot protection from slip stream and chemicals.
6. No hoses, valves, or any portion of the chemical carrying system should pass through the cockpit area, and no gas-pressurized systems should be used on the aircraft without proper design consideration and, usually, specific government inspection of the systems.
7. Levers, controls, and valves which the pilot has to operate during flight should be installed for ease in locating, identifying, and operating them. Frequently this is done by fitting the control handle or grip in such a way that it can be identified by touch.
8. It is to be specifically noted that all aircraft equipment installation or alteration in most countries requires specific approval of the actual work done by a licensed mechanic, and that request for approval of a specific installation must be made and permission granted by the regulatory agency before the aircraft can be legally flown.
9. Lastly, it is no doubt evident from the above list of basic aircraft requirements that only airplanes specifically designed and properly equipped for application work should be used. Converting a utility airplane for aerial application is sometimes done and may be justified in an emergency situation; however, in view of the significant hazards involved in aerial application work, the equipment and its installation should be the very best available and should follow the specific installation criteria generally applied to aeronautical equipment rather than those used for ground-operated chemical application equipment.

14. AGRICULTURAL PILOT TRAINING

Formal training schools for agricultural pilots and operators have increased in recent years with the need for replacements for the post-World War II pilots, who are now reaching retirement age or have gone into aerial-application business management (Jose, 1969). There has also been a steady increase in the demand for new pilots in most areas as the application business has expanded. Machines, methods, and materials have become more sophisticated, thus increasing the need for more formal training in the area, as well as for more basic flight training in low-elevation precision flying.

A large proportion of the present aerial-application operators and pilots were self-trained in agricultural flying, many having received basic flight training with a few hundred hours of flight time on military aircraft. The essential training for an agricultural pilot involves operating the aircraft and the application equipment in flight. The first generation of agricultural flyers not only had to teach themselves how to fly at low elevations and around various obstructions, but they also had to sell aerial application to a sceptical public, which required at least a rudimentary knowledge of the materials used and pests being controlled. Meanwhile, they had to manage to survive the very high accident rate which plagued the early agricultural aircraft business for many years.

The flight-training requirement for a student agricultural pilot is generally satisfied by holding a commercial pilot's licence — which means that in the U.S.A., for example, he has had a minimum of 160 hours of flight training (dual and solo) and sufficient ground-school training to pass a written examination. Flight-training schools require a commercial pilot's licence plus anywhere from 250 to 500 hours of flight time for entry into their agricultural programme. Government licensing requirements vary. As might be expected, a usual requirement is a minimum agricultural flight time of 50-100 hours, plus the passing of a written examination on techniques, responsibilities, and regulations governing agricultural work. Agricultural pilot schools will usually call for 40-50 hours of dual and solo agricultural flying plus 50-100 hours of classroom study with a minimal coverage of application techniques, equipment calibration, chemicals and pests, and local application regulations. Helicopter training for a commercial helicopter pilot's licence usually calls for at least 25 hours of flight training beyond the commercial pilot's licence for fixed-wing aircraft. Agricultural helicopter training certificates from approved schools can be earned with about the same hours of flight training as for the agricultural fixed-wing pilot, or about 50 hours beyond the CPL.

With the heavy emphasis and undeniable need for a strong flight-training programme in agricultural pilot's school, the ground classes, particularly with regard to proper place, time, amounts, and methods for applying chemicals and other materials, have of necessity taken second place. In the U.S.A. the basic types of training programmes outlined above cost from $2 500 to $3 000 for tuition only for a 3 to 5 month total course. If in-depth training is provided in basic physics, chemistry, entomology, and crop protection and production, then 500-600 class hours (40-60 semester hours) are required, along with two or two and a half years of schooling at an applied level or at a terminal type two-year junior college. Tuition for nonresident students can be $500-$1 500 per semester, which along with the basic flight-training raises the costs to $8 000-$12 000 for tuition only. Housing, food, and other necessities must be added to this.

While it can be said that this is the type of training needed as a minimum requirement for agricultural pilots, it has been found difficult, at least up to now, to persuade young ex-military pilots, even with a government educational subsidy, to enter these programmes. The route for most people entering agricultural aviation today is still through basic flight training, with either military or civilian instruction, and apprenticeship to an agricultural aircraft operator having a dual-control aircraft in which he can personally train the prospective agricultural pilot. Little formal classroom training is provided — generally just enough to pass the rudimentary examinations given by the local licensing authority. In many instances the local governmental authority will have specific regulations governing licensing, and apprenticeship under a licensed operator is frequently specified.

Under these circumstances the incentive for in-depth formal training is lacking, since the pilot is paid by the operator while he is learning and is also assured of a job (if he meets the qualifications decided by the operator) when he completes his apprenticeship.

Pilot training is not likely to change greatly unless stronger government regulations are initiated, specifying longer hours of formal classroom training (or requiring a detailed examination) in basic agricultural application methods and business operations. In fact, a logical position to take is that there is no real need for every pilot to have a two- to four-year college education to function adequately as a pilot. Probably the most important person to train is the agricultural aviation manager or operator, who must ultimately be responsible for the pilots and others who work for him. Most of the large government programmes employ biological specialists, entomologists, agronomists, and others who advise on the proper materials to use and are able to evaluate the results obtained. A few commercial operators have hired trained biologists, but more frequently the aircraft operators look to chemical company sales and technical representatives for advice on proper chemicals, application methods, and dosages. This has not always been a satisfactory arrangement, as sales people must of necessity sell their product in order to continue in business; however, in view of increased regulation and local-government licensing of sales and technical personnel who advise on materials and methods of use, there is hope that more responsible and careful recommendations will be made, which will help relieve aircraft operators and pilots of some responsibility and provide better and more factual recommendations for specific materials and their proper use.

15. SPECIFIC TREATMENT PRACTICES

The basic data on the types of aircraft and application equipment and its functions have been described in detail in the preceding text. It might be well, however, to detail briefly some specific crops and the types of aircraft and treatments that are used in normal operations involving these crops.

COTTON

As has been noted, the use of aircraft for agricultural work in the U.S.A. was initiated for and continues to be the principal means of applying materials to cotton. Dusts were used at first, but high drift losses and consequent lawsuits and complaints of neighbours forced the shift to spray materials. The basic spray systems were for coarse aerosols (Table 11) or for fine sprays. Although these produce much less drift hazard than dust, nevertheless such sprays, generated by small-orificed hydraulic nozzles under significant pressures and with wide spray angles or by spinning wire atomizers for a volume median drop size of 100-200 μm, still present significant airborne drift and losses of 25% up to 60% of active chemicals from the field being treated. While the use of coarser sprays, up to 500 μm VMD, will reduce these losses to 10-20%, there is a loss of plant coverage by the larger drops because these are fewer in numbers in a given volume of liquid. This has led to a very complex situation in relation to applied volumes of liquids. The low- and ultra-low-volume techniques reduce or eliminate the highly vaporizable water content of the spray formulation and thereby reduce drift losses by maintaining larger (initial) spray drop size. But, in order to obtain plant coverage or even airspace (volume) coverage as required for adulticiding of insects, drop size must be reduced as the volume is decreased. Since the individual drop volume is a d^3 (diameter cubed) function, the drop volume decreases rapidly as size is reduced, or the number of drops in a given volume of liquid (as shown in Table 7) increases (for a given volume) inversely to their diameter cubed.

For cotton-insect control the fine sprays offer the greatest possible coverage of the plants and also permit reduced applied-volume or low-volume treatment. But, unfortunately, losses from the fine sprays are very high, which tends to reduce the effectiveness of the treatment and more importantly permits widespread drift and damage to local insect ecology, affecting the predator and parasites of economic insects, as well as creating the danger of unwanted crop residues. While the change from chlorinated hydrocarbons to carbamate and phosphate chemicals has reduced the residue problem significantly, the damage to insect ecology has been greatly increased due to extreme sensitivity of predator and parasite insects to those chemicals. In many places in the world the economic insects of cotton have gotten out of control owing to the suppression of normal predators and parasites as well as to the development of chemical-resistant strains of insects. The present control trend is to hold off early treatments of cotton with early insect-control chemicals and reduce the number of treatments and amount of chemical used as a means of trying to protect favourable insect species.

The largest drop-size sprays and large granular materials (where possible) should be used for insect and fungus control and for herbicide applications where coarse sprays and granular materials are already well established because of the necessity of confining chemical herbicides to treated fields. Therefore, the basic means for reducing chemical drift is to use large particles of granular materials or sprays which settle rapidly and have but few particles under 50 μm size, which are prone to drifting. Additives such as cellulose, inverted phase-crowded emulsions, and foam systems reduce drift by the simple expedient of increasing average drop size. Except for certain low-turbulence nozzle systems, however, all atomizers and formulations produce a given number of fine drift-prone drops, so drift control cannot be totally achieved at this time.

Rice

Another crop on which aircraft are widely used the world over is rice. In the U.S.A., the U.S.S.R., and France relatively large rice fields make this a practical means for both fixed- and rotary-wing aircraft applying seed, fertilizers, herbicides, insecticides, and fungicides, as well as mosquito-control chemicals. But in most of the Asian countries, where the largest areas of rice production exist, small fields and interplanted crops make treatment by fixed-wing aircraft impractical if not hazardous, so rotary-wing aircraft are logically the better choice. However, the higher cost of helicopters keeps this machine from widespread use except where heavy economic subsidy is offered or government-sponsored (military or other) helicopters are used.

The application of fertilizers to wet rice fields requires a high-load capability in the aircraft and has led to the development of very large single-engine aircraft designed for applying high rates of 336-560 kg/ha (300-500 lb/acre; see Fig. 7). Although other applications to rice will not need such large-capacity machines, the tendency to use these for all operations promotes the use of large aircraft in rice areas.

ORCHARD AND VINE CROPS

In general, the use of aircraft has been very limited with respect to crops of significant height or depth or with heavy foliage, since penetration from above is not very practical and thorough coverage of the foliage in depth usually requires much lower volumes than with ground equipment. As with other crops, the tendency to try to obtain in-depth coverage by using coarse aerosol sprays or dusts has resulted in heavy drift losses and contamination problems. Moreover, the insect ecology of fruit, nut, and vine crops is very complex and highly dependent on predator and parasite insects. Thus, all of the complications inherent in cotton plant protection exist, plus the problem of in-depth coverage of tall trees and vines.

Nevertheless, there has been an increasing use of aircraft in this area. The helicopter is most favoured, particularly for applications during the dormant (winter) period of the season. Many insect, weed, and fungus control chemicals can be applied at this time, and fine sprays are less likely to damage other crops. While thorough control of insects and disease is difficult to achieve in an orchard or a vineyard in full foliage, rapid treatment potential and satisfactory economic control levels have made possible the use of helicopters to a limited extent even in the nondeciduous citrus plantings.

Uses of Aircraft for Large-Scale Programmes

Aircraft have from their inception been widely used, particularly in the U.S.S.R., the U.S.A., and the Scandinavian countries, for applications to forest and rangelands for insect and disease control, as well as more recently for fertilizer applications. Here, the aircraft used have frequently been multi-engined ex-military or cargo types designed for relatively high flight operation over large areas. As noted earlier under operational analysis (see page 121), large aircraft are, however, frequently more expensive to use per unit of area covered than smaller single-engine agricultural types. This is particularly true of the aircraft used for applying very low (or ultra-low) volumes of chemicals for vector and forest insect control. Again, the difficulty of obtaining a good deposit with aerosol applications leads to greater dependence on applications made under low-wind, temperature-inversion conditions and, of course, to widespread drift or to contamination of entire air-shed areas by the chemicals. While very low rates per acre may be used, the accumulative total and highly airborne character of the particle size used give rise to serious air contamination and, finally, wildlife ecology problems.

Application of Biological Materials

The problems of drift and contamination are basically related to the potential hazard of the released chemical. If it were possible to obtain economic control by releasing biological materials — virus and bacteria, as well as sterile insects, parasites and predators, or even sex pheromones (to confuse the normal mating) — which would have no ecological effect on the area, then the problems of ecological impact could be solved. But biological control, with a few exceptions, is not easily practised nor as effective and rapid as might be necessary in many plant protection cases. At present the use of biological materials is basically a supplemental technique, which aids the overall control programme and makes possible a lesser use of pesticide chemicals. To this end, all means, including biological ones, should be employed, and the system management or integrated control techniques seem to point the way to eventual fulfillment of our plant protection needs with considerably less dependence on chemicals.

FIRE-FIGHTING AIRCRAFT

The wide use of aircraft in the control of forest and brushland fires can provide the aircraft owner-operator with still another outlet for obtaining more hours of use for his aircraft. While description of the large aircraft and the systems used by these for fire fighting is outside the scope of this text, it should be pointed out that smaller agricultural aircraft do participate in fire fighting with little or no alteration of their basic spray system. Usually the spray boom is removed, and the emergency dump system is opened to drop 500-1 000 litres (130-260 U.S. gals) of water or fire-retardant mixture on the fire from overhead. Small planes and turbine-powered helicopters are particularly useful in hilly areas and on brushland, as a closer approach to the fire can aid in more accurate and lower-volume fire control.

Many other selected areas of special aircraft use which might be delineated are beyond the scope of this text. The ingenuity of aircraft operators will continue to develop new areas of use, and most likely aircraft are here to stay in agriculture, forestry, and other food and fibre production endeavours.

APPENDIX TABLES

TABLE 1. — WORLD-WIDE AIRCRAFT USE FOR PLANT SEEDING, PROTECTION, AND NUTRITION AS WELL AS VECTOR AND LOCUST CONTROL. (Treated area is actual hectares × number of treatments)

Country	Aircraft (numbers)	Treated hectares (thous.)	Country	Aircraft (numbers)	Treated hectares (thous.)
Algeria	7	42	Israel	33	467
Angola	4	10	Italy	32	264
Argentina	450	5 000	Japan	158	1 622
Australia	260	6 170	Madagascar	10	57
Austria	17	18	Malaysia	1	5
Bulgaria	80	160	Mexico	450	8 000
Cameroon	6	30	Morocco	12	36
Canada	666	2 130	Mozambique	2	8
Chile	20	127	Netherlands	18	70
China	200(est.)	1 600(est.)	New Zealand	213	3 320
Colombia	208	2 563	Nicaragua	190	3 834
Costa Rica	9	84	Norway	5	4
Cuba	184	5 152	Pakistan	50(est.)	41
Cyprus	1	4	Peru	170	1 000
Czechoslovakia	92	665	Philippines	10	50
Denmark	14	56	Poland	50(est.)	300
El Salvador	136	3 395	Portugal	10	38
Egypt	23	81	South Africa	30	3 000
Ethiopia	5	425	Spain	89	1 674
Finland	7	20	Sudan	50	1 300
France	50	169	Surinam	5	81
German Dem. Rep.	100	1 840	Sweden	36	180
Germany, Fed. Rep. of	18	44	Syrian Arab Rep.	6	25
Greece	15	523	Taiwan	(Limited)	
Guatemala	157	3 432	Thailand	6	
Honduras	23	470	Turkey	55	464
Hungary	34	320	United Kingdom	47	245
India	28	405	Uruguay	70	1 100
Indonesia	9	1 050	United States	6 100	42 100
Iran	37	550	U.S.S.R.	8 000	81 000
Iraq	14	51	Yugoslavia	94	1 200

TABLE 2. — TRENDS IN MATERIAL FORMS AND USES OF AIRCRAFT APPLICATIONS, AREAS TREATED, AND NUMBERS OF AIRCRAFT

U.S.A.	1950	1960	1970
	Percentage of total treated area (ha × no. of treatments)		
Forms:			
Spray pesticides	38	46	75 (*est.*)
Dust pesticides	49	39	5
Granular pesticides	—	3	8
Fertilizer	6	5.5	7
Seeds	7	6.5	5
Uses:			
Agr. defoliant	4	3.5	5 (*est.*)
Insecticide	—	73.3	59
Fungicide	—	1.8	8
Herbicide	—	12.7	18
Forest (insect)	—	2.7	3
Miscellaneous	5	6	7
Total hectares treated (millions)	16.2	29	42
Total number of aircraft	4 500	5 130	6 100

Yugoslavia	1962	1964	1969
	Percentage of total treated area (ha × no. of treatments)		
Forms:			
Spray and dust	32	26.6	36
Dust	<10	<10	4.4
Fertilizers	43	29	50
Forest & mosquito	25.2	43.7	9.4
Miscellaneous	0.1	0.7	0.2
Total hectares treated (millions)	0.123	0.414	1.2
Total number of aircraft	12	21	94

World-wide total use:			
Total hectares treated (millions)		60	157
Total number of aircraft		11 000	17 000

TABLE 3. — FIXED-WING AIRCRAFT. Some aircraft types available commercially now or in the recent past. Wing type: (*B*) biplane, (*L*) low-wing monoplane, (*H*) high-wing monoplane. Maximum horsepower ratings. Chemical load under restricted agricultural use. Run, climb, working speed, and stall at sea level with spray gear and under normal gross-weight conditions. Approximate cost classes ($U.S.): (*A*) over $50 000; (*B*) $25 000-50 000; (*C*) under $25 000 (with application equipment).

Mfg. and country	Model (wing type)	Horse-power	Gross weight (normal) Kg (lb)	Chemical load (restricted) Kg (lb)	Take-off run M (ft)	Rate of climb M/min (ft/min)	Working speed Km/hr (mi/hr)	Stall speed (flaps) Km/hr (mi/hr)	Cost class
Air Parts (N.Z.)	Fletcher FU-24 (*L*)	300	2 029 (4 470)	730 (1 600)	152 (500)	190 (625)	180 (112)	78 (48)	*B*
	FU-24-950 (*L*)	400	2 358 (5 200)	1 044 (2 300)	140 (460)	280 (920)	214 (133)	81 (50)	*B*
	Turbo-1060 (*L*)	500	2 472 (5 420)	1 361 (3 000)	256 (850)	290 (950)	177 (110)	88 (55)	*A*
Anahuac (Mexico)	El Tauro 300 (*L*)	300	1 606 (3 542)	800 (1 760)	— (—)	152 (500)	137 (85)	81 (50)	*C*
Antanov (U.S.S.R.)	An-2M (*B*)	1 000	5 500 (12 250)	1 960 (4 312)	200 (655)	132 (433)	200 (124)	75 (47)	*A*
Bellanca	Champion Scout (*H*)	150	1 054 (2 325)	363 (800)	— (—)	107 (350)	137 (85)	97 (60)	*C*
Canadair (Canada)	CL-215 (*H*)	2 100 (Twin)	16 329 (3 600)	4 900 (10 800)	151 (1 940)	290 (950)	280 (174)	145 (90)	*A*
Cessna (U.S.A.)	Ag Pickup (*L*)	230	1 497 (3 300)	757 (1 660)	341 (1 120)	121 (400)	145 (90)	92 (57)	*C*
	Ag Wagon (*L*)	285	1 497 (3 300)	757 (1 660)	257 (845)	210 (690)	183 (114)	92 (57)	*B*
	Ag Truck (*L*)	285	1 497 (3 300)	1 056 (2 332)	207 (680)	210 (690)	182 (114)	92 (57)	*B*
	Ag Carryall								

	Model											
(Australia)												
Dehavilland (Canada)	Chipmunk DHC-1 (L)	145	1 100	(2 420)	360	(794)	230	(750)	152	(500)	145	(90)	52	(84)	C
	Beaver DHC-2 MKI (H)	450	2 313	(5 100)	910	(2 000)	170	(560)	311	(1 020)	201	(125)	72	(45)	A
	Beaver Turbo DHC-2 MKIII (H)	578	2 313	(5 100)	910	(2 000)	152	(500)	361	(1 185)	225	(140)	97	(60)	A
	Otter DHC-3 (H)	600	2 010	(4 431)	—	(—)	192	(630)	198	(650)	195	(121)	93	(58)	A
Dinfia (Argentina)	IA-46 Ranquel (H)	150	1 160	(2 555)	365	(803)	277	(910)	144	(470)	155	(96)	60	(37)	—
	Super Ranquel (H)	180	1 260	(2 775)	465	(1 030)	280	(920)	207	(680)	170	(106)	57	(36)	—
	IA 51	180	1 260	(2 775)	420	(925)	232	(760)	216	(710)	195	(121)	90	(56)	—
	IA 53	235	1 525	(3 362)	499	(1 100)	187	(615)	230	(755)	160	(100)	80	(50)	—
Emair (U.S.A.)	Paymaster MAI (B)	600	3 180	(7 000)	1 360	(3 000)	—	(—)	152	(500)	185	(117)	95	(59)	A
Funk (U.S.A.)	F-23-B	275	1 950	(4 300)	680	(1 500)	260	(850)	156	(515)	161	(100)	92	(57)	C
Grumman (U.S.A.)	AgCat (B)	450	2 041	(4 500)	908	(2 000)	228	(750)	329	(1 089)	161	(100)	107	(67)	B
	Super AgCat (B)	600	2 041	(4 500)	1 133	(2 500)	120	(395)	—	(—)	177	(110)	—	(—)	B

TABLE 3. — FIXED-WING AIRCRAFT. Some aircraft types available commercially now or in the recent past. Wing type: (B) biplane, (L) low-wing monoplane, (H) high-wing monoplane. Maximum horsepower ratings. Chemical load under restricted agricultural use. Run, climb, working speed, and stall at sea level with spray gear and under normal gross-weight conditions. Approximate cost classes ($U.S.): (A) over $50 000; (B) $25 000–50 000; (C) under $25 000 (with application equipment). *(concluded)*

Mfg. and country	Model (wing type)	Horse-power	Gross weight (normal) Kg (lb)	Chemical load (restricted) Kg (lb)	Take-off run M (ft)	Rate of climb M/min (ft/min)	Working speed Km/hr (mi/hr)	Stall speed (flaps) Km/hr (mi/hr)	Cost class
IAR (Rumania)	IAR-822 (L)	290	1 900 (4 200)	600 (1 322)	170 (558)	210 (688)	185 (115)	75 (47)	—
North Am. Rockwell (U.S.A.)	Aero-Comm. Sparrow A-9 (L)	235	1 362 (3 000)	640 (1 400)	244 (800)	198 (650)	153 (95)	88 (55)	C
	Quail A-9B (L)	290	1 663 (3 600)	794 (1 750)	198 (650)	260 (850)	161 (100)	88 (55)	C
	Snipe B-1 (L)	450	2 041 (4 500)	908 (2 000)	183 (600)	198 (650)	161 (100)	73 (45)	B
	Thrush S-2R (L)	600	2 722 (6 000)	1 133 (2 500)	236 (775)	274 (900)	177 (110)	105 (66)	B
OKL (Poland)	Gawron (H) DZL-101-A	260	1 660 (3 670)	500 (1 103)	130 (426)	162 (530)	130 (81)	49 (30)	—
	Wilga 3R (H) DZL-104	260	1 150 (2 530)	272 (600)	90 (295)	510 (1 670)	190 (118)	70 (44)	—
Omnipol (Czec.)	Cmelak (L) Z-37	315	1 850 (4 080)	600 (1 323)	155 (509)	222 (726)	120 (75)	77 (48)	B
Pilatus (Swit.)	Porter PC-6 (H)	350	1 960 (4 322)	800 (1 750)	130 (296)	260 (850)	187 (116)	70 (44)	B

	B1-H2 (H)	550	2 200	(4 850)	1 176	(2 800)	198	(648)	460	(1 520)	202	(126)	112	(70)	A
Piper Aircraft (U.S.A.)	Super Cub PA13A (H)	150	949	(2 090)	370	(820)	92	(300)	232	(760)	145	(90)	69	(43)	C
	Pawnee C PA-25 (L)	235	1 315	(2 900)	635	(1 400)	244	(800)	192	(630)	169	(105)	98	(61)	B
	PA-25-L	260	1 315	(2 900)	635	(1 400)	207	(680)	215	(705)	171	(106)	98	(61)	B
	Pawnee Brave PA-36 (L)	285	1 970	(4 400)	860	(1 900)	488	(1 600)	107	(350)	145	(90)	106	(66)	B
Socata (France)	Agricorallye MS-893	180	1 050	(2 318)	347	(765)	160	(525)	210	(690)	217	(135)	92	(57)	—
Transavia (Australia)	Airtruk PL-12 (H)	300	1 854	(4 090)	822	(1 810)	334	(1 095)	183	(600)	175	(108)	95	(59)	B
UTVA (Yugoslavia)	U-600-AG (H)	270	1 730	(3 814)	545	(1 200)	195	(640)	372	(1 200)	219	(136)	66	(41)	—
	U-65	295	1 463	(3 225)	600	(1 323)	144	(473)	144	(474)	178	(96)	80	(43)	—
Wetherly (U.S.A.)	201-A (L)	450	2 176	(4 800)	908	(2 000)	335	(1 100)	292	(960)	169	(105)	111	(69)	B

Data: Manufacturers' specifications and International Agricultural Aviation Centre, The Hague.

TABLE 4. — ROTARY-WING AIRCRAFT. Some aircraft types which are or have been available commercially. Maximum horsepower ratings at sea level. Chemical load under restricted agriculture category. Rate of climb and working speed maximum at sea level and under normal gross weight without application equipment. Approximate cost classes ($U.S.): (A) over $50 000 and (B) under $50 000 (with application equipment).

Mfg. and country	Model	Horse-power	Gross weight (normal) Kg (lb)	Chemical load Kg (lb)	Rate of climb M/min (ft/min)	Working speed Km/hr (mi/hr)	Service ceiling (max) M (ft)	Cost class
Bell (Textron) (U.S.A.)	47G-3B1	270	1 340 (2 950)	362 (800)	269 (880)	128 (80)	6 400 (21 000)	B
	47G-3B2	280	1 340 (2 950)	454 (1 000)	302 (990)	141 (88)	5 610 (18 400)	A
Agusta (Italy)	47G-4A	280	1 340 (2 950)	454 (1 000)	244 (800)	136 (85)	3 415 (11 200)	A
	47AG5	265	1 290 (2 850)	544 (1 200)	262 (860)	135 (84)	3 200 (10 500)	B
Brantly (U.S.A.)	B2B	180	755 (1 665)	190 (420)	427 (1 400)	153 (95)	3 320 (10 900)	B
	305	305	1 315 (2 900)	390 (860)	297 (975)	169 (105)	3 048 (10 000)	A
Continental Copters (U.S.A.)	El Tomcat MKIIIB	235	1 180 (2 600)	510 (1 120)	302 (990)	141 (88)	5 610 (18 400)	B
Dornier (Germany, Fed. Rep. of)	DO-132	440	1 300 (2 860)	595 (1 310)	— (—)	185 (114)	6 100 (20 000)	—
Fairchild Hiller (U.S.A.)	UH 12E	305	1 247 (2 750)	576 (1 050)	463 (1 520)	145 (90)	4 756 (15 600)	B
	UH SL4	320	1 361 (3 000)	454 (1 000)	366 (1 200)	145 (90)	5 488 (18 000)	A
	FH-1100	317	1 247 (2 750)	500 (1 100)	488 (1 600)	217 (133)	4 878 (16 000)	A

Manufacturer	Model							
Hughes (U.S.A.)	300	180	757 (1 670)	317 (700)	350 (1 150)	97 (60)	3 963 (13 000)	B
	300-C	190	950 (2 100)	465 (1 025)	398 (1 305)	160 (99)	4 570 (15 000)	A
	500 (Turbo)	317	1 156 (2 550)	635 (1 400)	518 (1 700)	145 (90)	— (—)	A
Kamov (U.S.S.R.)	KA-26	325 (Twin)	3 160 (6 967)	800 (1765)	— (—)	100 (62)	3 000 (9 840)	—
Kawasaki (Japan)	KHA (Bell 47G-3B)	270	1 340 (2 950)	400 (880)	268 (880)	140 (87)	5 640 (18 500)	B
Mikhail (U.S.S.R.)	Mi-2 (V-2 Turbo)	400 (Twin)	3 550 (7 800)	700 (1 540)	— (—)	210 (130)	4 000 (13 120)	—
	Mi-4-S	1 700	7 800 (17 200)	1 000 (2 200)		160 (99)	5 500 (18 000)	—
Siai-Marchetti (Italy)	Silvercraft SH-4	235	860 (1 900)	200 (441)	350 (1 140)	128 (80)	4 000 (13 120)	B
Sud Aviation (France)	SO1220 Djinn	260	760 (1 765)	300 (660)	— (—)	100 (60)	3 100 (10 170)	B
	SA318C Alouette II	360	1 650 (3 630)	600 (1 320)	390 (1 280)	180 (112)	3 300 (10 800)	A
	SA-341	600	1 700 (3 750)	750 (1 660)	660 (2 170)	245 (152)	5 716 (18 750)	A

TABLE 5. — PARTICLE DENSITY AND DISTRIBUTION OF SAMPLE AEROSOL DUSTS

Dust type	Particle density	Bulk density (packed)		Particle distribution
	g/cm^3	g/cm^3	lb/ft^3	microns (μ)
Talc	2.6	0.992	62	93% < 30μ
				80 < 20
				62 < 10
				46 < 5
				16 < 1
Attapulgite	2.3	0.464	29	100% < 50μ
				88 < 35
				69 < 25
				42 < 15
				12 < 5
Prophyllite	2.7	0.912	57	99% < 74μ
				59 < 10
				27 < 5
				11 < 2.5

TABLE 6. — APPROXIMATE SETTLING RATES OF DRY MATERIAL WITH STANDARD SCREEN SIZES

Screen size		Average number of granules		Average number of granules at 11.2 kg/ha (10 lb/acre)	
U.S.A. mesh	Opening microns (μ)	Per pound	Per gram	Per ft^2	Per m^2
8/16	2 360/1 000	145 × 10^3	0.32 × 10^3	30	323
16/30	1 000/ 520	1 150 × 10^3	2.5 × 10^3	260	2 800
18/35	749/ 417	2 708 × 10^3	6.0 × 10^3	620	6 674
25/50	673/ 283	7 722 × 10^3	17.0 × 10^3	1 770	19 050
30/60	520/ 246	12 500 × 10^3	27.0 × 10^3	2 820	30 355

TABLE 7. — TERMINAL VELOCITIES OF WATER DROPS IN STILL AIR AND NUMBER PER GIVEN VOLUME IN RELATION TO UNIT AREA AND AIR VOLUME

Drop diameter	Terminal or steady state velocity		Number of drops at applied rate of 9.35 litres per hectare (1 gal/acre)		
			Surface		In air to depth of 20 m (65.5 ft)
microns (μ)	ft/sec	m/sec	per in^2	per cm^2	per cm^3 (air)
1	$.01 \times 10^{-2}$	$.033 \times 10^{-3}$	$1\,153 \times 10^6$	180×10^6	900×10^2
5	$.25 \times 10^{-2}$	$.76 \times 10^{-3}$	91.8×10^6	1.4×10^6	7.0×10^2
10	1.0×10^{-2}	3.0×10^{-3}	1.15×10^6	0.8×10^6	0.9×10^2
20	4.0×10^{-2}	12.0×10^{-3}	0.14×10^6	0.022×10^6	10.9×10^2
50	0.25	0.076	91.0×10^2	14.0×10^2	0.64
70	0.4	0.12	34.0×10^2	5.3×10^2	0.27
100	0.9	0.27	11.5×10^2	1.8×10^2	0.09
150	1.5	0.46	3.5×10^2	0.53×10^2	—
200	2.5	0.76	1.4×10^2	0.22×10^2	—
					per ft^2 (surface)
300	3.8	1.15	43.0	6.7	6 192
500	7.0	2.13	9.0	1.4	1 311
1 500	13.0	4.0	1.13	0.174	162
2 000	21.0	6.4	0.14	0.022	20.2
5 000	30.0	9.2	0.09	0.009	13.0

NOTE: The above data apply specifically to water drops; solid particles have approximately the same characteristics, depending on density.

TABLE 8. — PHYSICAL PROPERTIES OF SELECTED LIQUID MATERIALS

Liquid	Surface tension dynes/cm (20°C)	Density g/cm^3	Viscosity cP at 20°C	Vapour pressure mmHg at	temp. °C
Acetone	24	.79	0.32	195	20
Methanol	22	.8	0.6	100	20
Benzene	30	.9	0.65	80	20
Xylene	30	—	0.68	10	32
Water	73	1.0	1.0	55	40
Water	72.8	1.0	1.0	18	20
Ethanol	22	.79	1.2	47	20
Gasoline	—	.68	0.35	—	—
Turpentine	—	.867	1.49	3	20
Kerosene	25	.82	2.5	7	30
Diesel fuel	30	.89	10	—	—
Ethylene glycol	47	—	20	—	—
Propylene glycol	—	—	—	1	45
Cottonseed oil	35.4	.92	70	—	—
Lube oil SAE10	36	.9	100	—	—
Lube oil SAE30	36	.9	300	1	30
Glycerol	63	1.26	800	1	125
Castor oil	39	.97	1 000	—	—
Corn syrup	78	—	10 000	—	—
Malathion (95%)	32	1.23	45	4.0×10^{-5}	30
Lindane	—	—	—	9.4×10^{-6}	20
DDT	—	1.4(solid)	—	1.9×10^{-7}	20
Parathion (ethyl)	—	1.35	—	4.0×10^{-5}	20
2,4-D (isopropylester)	—	—	—	10.5×10^{-3}	25
Dursban (75%)	—	.97	—	1.87×10^{-5}	25
Naled (85%)	—	1.965	—	—	—
Fenthion (93%)	—	1.25	—	2.15×10^{-6}	20
Captan	—	1.73	—	1.0×10^{-5}	25

Non-Newtonian thickening agents	Viscosity cP at 20°C, showing two rates of shear	
	1/50 sec	1/4 000 sec
Invert emulsion: 5% emulsifier, 10% diesel fuel, 85% water	700	16
Remaining mixtures each contain 2.5% Tordon (herbicide) in total of 100 gals water and mix		
Norbak 6.2 lb	680	45
Vistik 6.5 lb	450	16
Keltex 8.5 lb	260	39
Dacagin 6.0 lb	140	13

From Butler, Akesson, and Yates, 1969.

Table 9. — Twin-fluid air/liquid atomizers: Spraying System Co. nozzle type 10900J fluid, petroleum oil, 45 centipoise

Liquid orifice*	Liquid pressure		Flow rate		Air pressure		Flow rate		VMD
	kg/cm^2	lbf/in^2	l/min	gal/min	kg/cm^2	lbf/in^2	m^3/min	ft^3/min	micrometres
1050	.07	1	.01	.0026	.35	5	.026	.93	15 µm
1050	.35	5	.022	.0057	.35	5	.026	.93	25
1050	.7	10	.03	.0078	.35	5	.026	.93	35
1650	2.8	40	.18	.047	1.4	20	.045	1.6	85
2850	.7	10	.27	.07	1.4	21	.045	.6	96
2850	1.4	20	.38	.1	2.8	40	.065	2.3	103
2850	1.4	20	.38	.1	2.1	30	.057	2.0	120
2850	2.8	40	.53	.14	4.2	60	.079	2.8	88

* Liquid orifice diameters in inches approximately .01 (1050), .016 (1650), and .28 (2850); millimetres approximately .254, .406 and .71, respectively. Air orifice size remains the same, and air is measured at standard conditions or as free air.

Table 10. — Drop size and air velocity with water at 20°C (68°F)

Drop to air velocity		Maximum drop size at critical velocity	Maximum drop size at steady falling velocity
km/hr	mi/hr	micrometres	micrometres
80	50	1 500 µm	2 600 µm
105	65	900	1 500
137	85	535	900
161	100	385	650
241	150	170	300
322	200	100	160

TABLE 11. — AIRCRAFT SPRAY DROP SIZE RANGE, APPROXIMATE RECOVERIES, AND USES

Spray descriptions and atomizers	*Selected atomizers	**Drop-size ranges in micrometres (μm) VMD	***Percent est. deposit in 305 m (1 000 ft)	General uses
Fine sprays: Cone and fan nozzles, and rotary atomizers	80005 down D6-45 down (50-100 lbf/in²)	100-300	40-80	Primarily for forest pesticide chemicals and large-area vector control with low dosages of low-toxicity and rapid-degradation chemicals. Also for agricultural insect pathogens.
Medium sprays: Cone and fan nozzles, and rotary atomizers	8004 down D6-46 down (30-50 lbf/in²)	300-400	70-90	Commonly used spray drop size for all low-toxicity agricultural chemicals where good coverage is necessary.
Coarse sprays: Cone and fan nozzles, and spray additives	8004 back D6-46 back (39-50 lbf/in²)	400-600 with additives (up to 2 000)	85-98	Recommended for toxic pesticides of restricted classification where thorough plant coverage is not essential.
Minimum-drift sprays: Jet nozzles and spray additives	D4 to D8 down at less than 60 mi/hr; back for over 60 mi/hr (30-50 lbf/in²)	800-1 000 with additives	95-98	Recommended for all toxic, restricted herbicides, such as phenoxy acids, within the limitations of the growing season and nearness to susceptible crops.
Maximum drift control: Low-turbulence nozzles	Microfoil ® at less than 60 mi/hr airstream (Registered trademark of AmChem Corp., U.S.A.)	800-1 000	99+	Actual drift tests show one quarter the drift residue levels at 152 m (500 ft) down-wind from the Microfoil® compared with D4 to D8 jets used with restricted nonvolatile herbicides, phenoxy acids, and others in the area of susceptible crops, but subject to limitations of growing season and crop.

* Numbers refer to Spraying Systems Co. nozzles and down or back refer to position on aircraft boom.
** Drop size as determined with water-base sprays; oils would give smaller drops.
*** Deposit estimated in 305 m downwind. Weather conditions: wind velocity 5-8 km/hr; neutral temperature gradient. Material released under 3.05 m height.

TABLE 12. — LIFETIME AND FALL DISTANCE AT TERMINAL VELOCITY FOR WATER DROPLETS. Air temperature 30°C (86°F), RH 50 %, $\Delta Td_b - w_b = 7.7°C$ (45°F)

Drop diameter	Lifetime	Distance drop would fall in lifetime in still air (approx.)	
micrometres	*seconds*	*m*	*ft*
200	64.8	38.4	126
100	16.2	2.44	7.99
80	10.4	0.994	3.26
		cm	*in*
50	4.0	7.62	3
40	2.6	6.25	2.46
20	0.65	<2.54	<1
10	0.16	<2.54	<1
2	0.065	<2.54	<1

TABLE 13. — COST OF OPERATING AGRICULTURAL AIRCRAFT

		Aircraft cost class		
		*A	B	C
Fixed costs per year				
Depreciation	**(C - S)/10	$6 300	$3 150	$1 800
Interest	$\frac{C-S}{2} + S\,(10\%)$	3 850	1 925	1 100
Taxes and licence	$\frac{C-S}{2} + S\,(6\%)$	2 310	1 155	660
Hangar and airstrip	$\frac{C-S}{2} + S\,(3\%)$	1 155	578	330
Insurance (*see* Table 14)		4 738	3 295	2 675
Total fixed costs per year		$18 353	$10 103	$6 565
Variable costs per hour				
Fuel and oil (petrol 60 cents/gal)		$12	$9	$6
Repairs, inspection per 1 000 hours		18	16	14
Major engine overhaul per 1 000 hours		5	4	3
Pilot allowance		15	15	15
Total variable costs per hour		$50	$44	$38

Total costs per hour for specified hours of use by aircraft cost class

Hrs/yr	Fixed costs/hr			Variable costs/hr			total cost/hr		
	(A)	(B)	(C)	(A)	(B)	(C)	(A)	(B)	(C)
150	$122.35	$67.35	$43.77	$50	$44	$38	$172.35	$111.35	$81.77
300	61.18	33.68	21.88	50	44	38	111.18	77.68	59.88
600	30.59	16.84	10.94	50	44	38	80.59	60.84	48.94
1 200	15.29	8.42	5.47	50	44	38	65.29	52.42	43.47

* Cost class ($U.S.): (A) $70 000, (B) $35 000, (C) $20 000.
** C, initial cost; S, salvage value or 10% of C.

TABLE 14. — INSURANCE SCHEDULE

Coverage	Cost per aircraft per year		
	*A	B	C
(1) Fire, theft, damage on ground** $\frac{C-S}{2}$ + S (2.5%)	$963	$482	$275
(2) Crash or hull coverage	1 925	963	550
(3) Bodily injury and property damage (excluding chemical drift or misapplication), $250 000 max.	350	350	350
(4) Property or crop damage from misapplication or chemical drift. Costs are variable, depending on the hazards involved and the extent of the local damage problem. For all chemical-liability coverage, $100 000 max.	850	850	850
(5) Workman's compensation, $1 000 per pilot***	500	500	500
(6) Personal accident, hospitalization, $300 per pilot***	150	150	150
Total	$4 738	$3 295	$2 675

* Cost class ($U.S.): (A) $70 000, (B) $35 000, (C) $20 000.
** C, initial cost; S, salvage value or 10% of C.
*** Estimated on the basis of two aircraft per pilot.

TABLE 15. — OPERATIONS ANALYSIS DATA

Aircraft and applied rate	Payload	Rate	Load time	Ferry	Ferry speed	Run	Run speed	Swath	Turn time
	kg (lb)	kg/ha (lb/acre)	min	km (mi)	km/hr (mi/hr)	km (mi)	km/hr (mi/hr)	m (ft)	min
Fixed wing									
Heavy (see Figs. 66, 72).....	1 134 (2 500)	280 (250)	2	1.61 (1)	144.8 (90)	1.61 (1)	144.8 (90)	7.62 (25)	0.5
Normal (see Figs. 67, 73).....	907 (2 000)	112 (100)	2	3.22 (2)	144.8 (90)	1.61 (1)	144.8 (90)	12.19 (40)	0.5
Light (see Figs. 68, 74).....	680 (1 500)	0.56 (0.50)	5	16.09 (10)	144.8 (90)	8.05 (5)	144.8 (90)	91.44 (300)	0.5
Rotary wing									
Heavy (see Figs. 69, 75).....	544 (1 200)	280 (250)	2	0.8 (0.5)	128.7 (80)	1.61 (1)	96.6 (60)	7.62 (25)	0.25
Normal (see Figs. 70, 76).....	454 (1 000)	28 (25)	3	0.4 (0.25)	128.7 (80)	0.8 (0.5)	96.6 (60)	9.14 (30)	0.25
Light (see Figs. 71, 77).....	272 (600)	0.56 (0.50)	5	8.05 (5)	128.7 (80)	8.05 (5)	96.6 (60)	91.44 (300)	0.25

TABLE 16. — APPROXIMATE COSTS PER HECTARE (ACRE) FOR VARIOUS AIRCRAFT APPLICATIONS

Material form and particle size		Swath width		Volume applied		Productivity		*Cost $U.S. for application only	
		m	(ft)	per ha	(per acre)	ha/hr	(acres/hr)	per ha	(per acre)
Liquid, microns VMD									
**Aerosols	under 50	305	(1 000)	73 ml	(1 oz)	4 050	(10 000)	$.025–.05	($.01–.02)
**Fine Spray	75– 125	183	(600)	146 ml	(2 oz)	2 024	(5 000)	$.05 –.075	($.02–.03)
**Medium Spray	150– 200	61	(200)	730 ml	(10 oz)	405	(1 000)	$.125–.15	($.05–.06)
**Medium Spray	250– 350	30	(100)	1.17 l	(1 pt)	202	(500)	$.25 –.37	($.10–.15)
**Medium Spray	250– 350	30	(100)	9.35 l	(1 gal)	121	(300)	$.50 –.75	($.20–.30)
**Medium Spray	250– 350	30	(100)	28 l	(3 gals)	41	(100)	$2.50	($1.00)
**Medium Spray	250– 350	15	(50)	187 l	(20 gals)	20	(50)	$15.00	($6.00)
Coarse Spray	400–500	12	(40)	46.8 l	(5 gals)	41	(100)	$3.75	($1.50)
Spray Jets	600–900	11	(35)	93.5 l	(10 gals)	30	(75)	$5.00	($2.00)
Thick Sprays	1 000–5 000	11	(35)	234 l	(25 gals)	12	(30)	$20.00	($8.00)
Dry materials, microns									
Dusts	under 25	30	(100)	33.7 kg	(30 lb)	61	(150)	$5.00	($2.00)
**Granules	250-500	23	(75)	2.24 kg	(2 lb)	81	(200)	$1.25	($.50)
Granules	250-500	23	(75)	22.4 kg	(20 lb)	61	(150)	$5.00	($2.00)
Seeds	1 000-5 000	18	(60)	84 kg	(75 lb)	40	(100)	$6.20	($2.50)
Granules	1 000-5 000	9	(30)	337 kg	(300 lb)	30	(75)	$7.40	($3.00)

* Costs are approximate for single-engine fixed-wing aircraft. Helicopters may cost somewhat more, but are frequently competitive where the work load is high.
** Items marked with asterisks are for large-area contract operations; others are for smaller operations, ranging from a few hectares to 200-300 ha (490-750 acres).

TABLE 17. — CHARACTERISTIC FLOW RATES OF SPRAYING SYSTEM

SPRAY ANGLE AT 40 p.s.i.	NOZZLE NO. TYPE T FEMALE CONN.	NOZZLE NO. TYPE TT MALE CONN.	"A"	Approx. Equiv. Orifice Dia.	"E"	MESH OF SCREEN	CAPACITY G.P.M. (Gallons Per Minute) AND SPRAY ANGLE AT p.s.i. (Lbs. per sq. inch) 5	7	10	15	20	30	40	60	80	100	150	200	300	40[0]
50°	• 1/4T 50 0017	1/4TT 50 0017	13/64"	.011	.007	200					.012 27°	.015 42°	.017 50°	.021 59°	.024 65°	.027 69°	.033 72°	.038 74°	.047 78°	.05 81
	• 1/4T 50 0025	1/4TT 50 0025	13/64"	.013	.009	200					.018 29°	.022 43°	.025 50°	.031 58°	.035 64°	.040 67°	.048 69°	.056 71°	.069 74°	.07 76
	• 1/4T 50 0033	1/4TT 50 0033	13/64"	.015	.011	200					.023 30°	.029 43°	.033 50°	.040 57°	.047 62°	.052 64°	.064 66°	.074 68°	.090 70°	.11 72
	• 1/4T 50 0050	1/4TT 50 0050	13/64"	.018	.013	200				.030 24°	.035 32°	.043 44°	.050 50°	.06 56°	.07 60°	.08 62°	.10 64°	.11 66°	.14 67°	.16 68
	• 1/4T 50 0067	1/4TT 50 0067	13/64"	.021	.014	100				.04 26°	.05 35°	.06 45°	.067 50°	.08 56°	.09 60°	.11 62°	.13 64°	.15 66°	.18 67°	.2 68
	1/4T 5001	1/4TT 5001	13/64"	.026	.018	100			.05 18°	.06 29°	.07 37°	.09 45°	.10 50°	.12 55°	.14 59°	.16 61°	.19 63°	.22 65°	.27 66°	.3 67
	1/4T 50015	1/4TT 50015	13/64"	.031	.022	100			.07 20°	.09 31°	.11 38°	.13 46°	.15 50°	.18 55°	.21 58°	.24 60°	.29 62°	.34 64°	.41 65°	.48 66
	1/4T 5002	1/4TT 5002	13/64"	.036	.024	50		.08 17°	.10 24°	.12 33°	.14 39°	.17 46°	.20 50°	.25 54°	.28 57°	.32 59°	.39 61°	.45 63°	.55 64°	.6 65
	1/4T 5003	1/4TT 5003	13/64"	.043	.033	50	.10 17°	.12 19°	.15 28°	.18 35°	.21 40°	.26 47°	.30 50°	.37 54°	.42 56°	.47 58°	.58 60°	.67 62°	.82 63°	.9 64
	1/4T 5004	1/4TT 5004	13/64"	.052	.037	50	.14 20°	.17 23°	.20 30°	.25 38°	.28 42°	.35 48°	.40 50°	.49 54°	.57 56°	.63 57°	.78 59°	.90 61°	1.10 63°	1.2 64
	1/4T 5006	1/4TT 5006	7/32"	.062	.044	50	.21 22°	.25 28°	.30 33°	.37 38°	.42 45°	.52 48°	.60 50°	.73 54°	.85 56°	.95 57°	1.16 58°	1.34 60°	1.64 62°	1.9 63
	1/4T 5008	1/4TT 5008	17/64"	.072	.054	50	.28 24°	.33 30°	.40 35°	.49 39°	.56 45°	.69 48°	.80 50°	.98 53°	1.13 55°	1.26 56°	1.55 57°	1.79 60°	2.20 62°	2.5 63
	1/4T 5010	1/4TT 5010	3/32"	5/64"	.063	*	.35 26°	.43 30°	.50 36°	.61 39°	.70 45°	.86 48°	1.0 50°	1.22 53°	1.41 55°	1.58 56°	1.93 57°	2.24 59°	2.74 61°	3.1 62
	1/4T 5015	1/4TT 5015	3/32"	3/32"	.071	*	.53 27°	.63 31°	.75 36°	.92 40°	1.16 48°	1.30 48°	1.5 50°	1.85 53°	2.12 55°	2.37 56°	2.90 57°	3.35 59°	4.10 61°	4.7 62
	1/4T 5020	1/4TT 5020	13/64"	7/64"	.083	*	.71 28°	.84 31°	1.0 36°	1.22 40°	1.41 45°	1.73 48°	2.0 50°	2.45 53°	2.82 55°	3.16 56°	3.87 57°	4.47 59°	5.48 61°	6.5 62
40°	• 1/4T 40 0017	1/4TT 40 0017	13/64"	.011	.007	200					.012 21°	.015 31°	.017 40°	.021 49°	.024 54°	.027 56°	.033 59°	.038 61°	.047 63°	.05 67
	• 1/4T 40 0025	1/4TT 40 0025	13/64"	.013	.009	200					.018 22°	.022 32°	.025 40°	.031 48°	.035 53°	.040 55°	.048 58°	.056 60°	.069 62°	.07 65
	• 1/4T 40 0033	1/4TT 40 0033	13/64"	.015	.011	200					.023 22°	.029 32°	.033 40°	.040 48°	.047 53°	.052 55°	.064 58°	.074 60°	.090 61°	.11 63
	• 1/4T 40 0050	1/4TT 40 0050	13/64"	.018	.013	200					.035 22°	.04 32°	.050 40°	.06 48°	.07 53°	.08 55°	.10 58°	.11 60°	.14 61°	.16 62
	• 1/4T 40 0067	1/4TT 40 0067	13/64"	.021	.015	100				.04 20°	.05 24°	.06 34°	.067 40°	.08 48°	.09 53°	.11 55°	.13 58°	.15 60°	.18 61°	.22 62
	1/4T 4001	1/4TT 4001	13/64"	.026	.018	100				.06 21°	.07 26°	.09 34°	.10 40°	.12 47°	.14 52°	.16 54°	.19 57°	.22 59°	.27 60°	.32 61
	1/4T 40015	1/4TT 40015	13/64"	.031	.020	100				.09 22°	.11 27°	.13 35°	.15 40°	.18 47°	.21 52°	.24 54°	.29 57°	.34 59°	.41 60°	.48 61
	1/4T 4002	1/4TT 4002	13/64"	.036	.026	50			.10 15°	.12 23°	.14 29°	.17 36°	.20 40°	.25 47°	.28 51°	.32 54°	.39 56°	.45 58°	.55 59°	.64 60
	1/4T 4003	1/4TT 4003	13/64"	.043	.031	50		.12 16°	.15 20°	.18 25°	.21 30°	.26 36°	.30 40°	.37 46°	.42 50°	.47 53°	.58 55°	.67 57°	.82 58°	.96 59
	1/4T 4004	1/4TT 4004	13/64"	.052	.038	50		.17 18°	.20 21°	.25 26°	.28 30°	.35 36°	.40 40°	.49 46°	.57 50°	.63 52°	.78 54°	.90 56°	1.10 57°	1.2 58
	1/4T 4006	1/4TT 4006	7/32"	.062	.048	50		.25 19°	.30 22°	.37 27°	.42 31°	.52 37°	.60 40°	.73 45°	.85 49°	.95 51°	1.16 53°	1.34 55°	1.64 56°	1.9 57
	1/4T 4008	1/4TT 4008	17/64"	.072	.057	50	.28 14°	.33 20°	.40 22°	.49 28°	.56 31°	.69 37°	.80 40°	.98 44°	1.13 47°	1.26 50°	1.55 52°	1.79 53°	2.20 54°	2.5 55
	1/4T 4010	1/4TT 4010	3/32"	5/64"	.065	*	.35 16°	.43 20°	.50 22°	.61 29°	.70 32°	.86 38°	1.0 40°	1.22 43°	1.41 45°	1.58 46°	1.93 47°	2.24 48°	2.74 49°	3.1 50
	1/4T 4015	1/4TT 4015	3/32"	3/32"	.076	*	.53 17°	.63 20°	.75 22°	.92 30°	1.16 32°	1.30 38°	1.5 40°	1.83 43°	2.12 45°	2.37 46°	2.90 47°	3.35 48°	4.10 49°	4.7 50
	1/4T 4020	1/4TT 4020	13/64"	7/64"	.089	*	.71 18°	.84 20°	1.0 24°	1.22 30°	1.41 32°	1.73 38°	2.0 40°	2.45 43°	2.82 45°	3.16 46°	3.87 47°	4.47 48°	5.48 49°	6.3 50
25°	• 1/4T 25 0017	1/4TT 25 0017	13/64"	.011	.008	200						.015 17°	.017 25°	.021 32°	.024 35°	.027 37°	.033 43°	.038 47°	.047 48°	.05 49
	• 1/4T 25 0025	1/4TT 25 0025	13/64"	.013	.010	200						.022 17°	.025 25°	.031 32°	.035 35°	.040 37°	.048 42°	.056 45°	.069 47°	.07 48
	• 1/4T 25 0033	1/4TT 25 0033	13/64"	.015	.012	200						.029 18°	.033 25°	.040 31°	.047 34°	.052 36°	.064 42°	.074 44°	.090 46°	.11 47
	• 1/4T 25 0050	1/4TT 25 0050	13/64"	.018	.015	200						.04 18°	.050 25°	.06 31°	.07 34°	.08 36°	.10 41°	.11 43°	.14 45°	.16 46
	• 1/4T 25 0067	1/4TT 25 0067	13/64"	.021	.017	100						.06 19°	.067 25°	.08 31°	.09 34°	.11 36°	.13 40°	.15 42°	.18 44°	.22 45
	1/4T 25 01	1/4TT 25 01	13/64"	.026	.020	100					.07 14°	.09 20°	.10 25°	.12 31°	.14 34°	.16 36°	.19 40°	.22 42°	.27 43°	.32 44
	1/4T 25 015	1/4TT 25 015	13/64"	.031	.025	100				.09 10°	.11 15°	.13 21°	.15 25°	.18 31°	.21 34°	.24 36°	.29 39°	.34 41°	.41 42°	.48 43
	1/4T 25 02	1/4TT 25 02	13/64"	.036	.029	50				.12 10°	.14 15°	.17 21°	.20 25°	.25 30°	.28 33°	.32 35°	.39 38°	.45 40°	.55 41°	.64 42
	1/4T 25 03	1/4TT 25 03	13/64"	.043	.037	50				.18 11°	.21 15°	.26 22°	.30 25°	.37 30°	.42 33°	.47 35°	.58 38°	.67 40°	.82 41°	.96 42

Co. HYDRAULIC NOZZLES OPERATED WITH WATER AT GIVEN PRESSURES

SPRAY ANGLE AT 40 p.s.i.	NOZZLE NO. TYPE T FEMALE CONN.	NOZZLE NO. TYPE TT MALE CONN.	"A"	Approx. Equiv. Orifice Dia.	"E"	MESH OF SCREEN	CAPACITY G.P.M. (Gallons Per Minute) AND SPRAY ANGLE AT p.s.i. (Lbs. per sq. inch) 5	7	10	15	20	30	40	60	80	100	150	200	300	400
	1/4T 25 04	1/4TT 25 04	13/64"	.052	.043	50			.20 5°	.25 12°	.28 16°	.35 22°	.40 25°	.49 29°	.57 32°	.63 34°	.78 37°	.90 39°	1.10 40°	1.26 41°
	1/4T 25 06	1/4TT 25 06	7/32"	.062	.054	50			.30 7°	.37 13°	.42 17°	.52 22°	.60 25°	.73 28°	.85 31°	.95 33°	1.16 36°	1.34 38°	1.64 39°	1.90 40°
	1/4T 25 08	1/4TT 25 08	17/64"	.072	.060	50			.40 8°	.49 14°	.56 17°	.69 22°	.80 25°	.98 29°	1.13 31°	1.26 33°	1.55 36°	1.79 38°	2.20 39°	2.52 40°
25°	1/4T 25 10	1/4TT 25 10	3/32"	5/64"	.068	*		.43 7°	.50 9°	.61 15°	.70 18°	.86 22°	1.0 25°	1.22 29°	1.41 31°	1.58 33°	1.93 36°	2.24 37°	2.74 38°	3.16 39°
	1/4T 25 15	1/4TT 25 15	3/32"	3/32"	.084	*		.63 8°	.75 10°	.92 15°	1.16 18°	1.30 22°	1.5 25°	1.83 29°	2.12 31°	2.37 33°	2.90 36°	3.35 37°	4.10 38°	4.75 39°
	1/4T 25 20	1/4TT 25 20	13/64"	7/64"	.094	*		.84 9°	1.0 11°	1.22 16°	1.41 19°	1.73 23°	2.0 25°	2.45 29°	2.82 31°	3.16 33°	3.87 36°	4.47 37°	5.48 38°	6.32 39°
•	1/4T 15 0017	1/4TT 15 0017	13/64"	.011	.008	200						.015 8°	.017 15°	.021 25°	.024 30°	.027 33°	.033 36°	.038 37°	.047 38°	.054 39°
•	1/4T 15 0025	1/4TT 15 0025	13/64"	.013	.010	200						.022 9°	.025 15°	.031 24°	.035 28°	.040 31°	.048 33°	.056 34°	.069 35°	.079 36°
•	1/4T 15 0033	1/4TT 15 0033	13/64"	.015	.012	200						.029 9°	.033 15°	.040 23°	.047 27°	.052 29°	.064 31°	.074 32°	.090 33°	.11 34°
•	1/4T 15 0050	1/4TT 15 0050	13/64"	.018	.015	200						.04 10°	.050 15°	.06 22°	.07 26°	.08 27°	.10 29°	.11 30°	.14 32°	.16 33°
•	1/4T 15 0067	1/4TT 15 0067	13/64"	.021	.017	100						.06 10°	.067 15°	.08 21°	.09 25°	.11 26°	.13 28°	.15 29°	.18 31°	.22 32°
	1/4T 15 01	1/4TT 15 01	13/64"	.026	.020	100						.09 10°	.10 15°	.12 21°	.14 24°	.16 25°	.19 27°	.22 28°	.27 30°	.32 31°
	1/4T 15 015	1/4TT 15 015	13/64"	.031	.025	100						.13 11°	.15 15°	.18 21°	.21 23°	.24 24°	.29 26°	.34 27°	.41 29°	.48 30°
15°	1/4T 15 02	1/4TT 15 02	13/64"	.036	.030	50					.14 6°	.17 11°	.20 15°	.25 20°	.28 22°	.32 23°	.39 25°	.45 27°	.55 28°	.64 29°
	1/4T 15 03	1/4TT 15 03	13/64"	.043	.038	50					.21 6°	.26 11°	.30 15°	.37 20°	.42 22°	.47 23°	.58 25°	.67 27°	.82 28°	.96 29°
	1/4T 15 04	1/4TT 15 04	13/64"	.052	.046	50					.28 7°	.35 12°	.40 15°	.49 19°	.57 21°	.63 22°	.78 24°	.90 26°	1.10 27°	1.26 28°
	1/4T 15 06	1/4TT 15 06	7/32"	.062	.056	50				.37 7°	.42 8°	.52 12°	.60 15°	.73 19°	.85 21°	.95 22°	1.16 24°	1.34 26°	1.64 27°	1.90 28°
	1/4T 15 08	1/4TT 15 08	17/64"	.072	.063	50				.49 7°	.56 9°	.69 13°	.80 15°	.98 19°	1.13 20°	1.26 21°	1.55 23°	1.79 25°	2.20 26°	2.52 27°
	1/4T 15 10	1/4TT 15 10	3/32"	5/64"	.071	*				.61 7°	.70 10°	.86 13°	1.0 15°	1.22 18°	1.41 19°	1.58 20°	1.93 22°	2.24 24°	2.74 25°	3.16 26°
	1/4T 15 15	1/4TT 15 15	3/32"	3/32"	.090	*			.75 6°	.92 7°	1.16 10°	1.30 13°	1.5 15°	1.83 18°	2.12 19°	2.37 20°	2.90 22°	3.35 24°	4.10 25°	4.75 26°
	1/4T 15 20	1/4TT 15 20	13/64"	7/64"	.102	*			1.0 6°	1.22 9°	1.41 10°	1.73 13°	2.0 15°	2.45 18°	2.82 19°	3.16 20°	3.87 21°	4.47 23°	5.48 24°	6.32 25°
	1/4T 00 0050	1/4TT 00 0050	9/64"	.018	.018	200	.018	.021	.025	.030	.035	.04	.050 0°	.06	.07	.08	.10	.11	.14	.16
	1/4T 00 0067	1/4TT 00 0067	9/64"	.021	.021	100	.024	.028	.033	.04	.05	.06	.067 0°	.08	.09	.11	.13	.15	.18	.22
	1/4T 00 01	1/4TT 00 01	9/64"	.026	.026	50	.035	.04	.05	.06	.07	.09	.10 0°	.12	.14	.16	.19	.22	.27	.32
	1/4T 00 015	1/4TT 00 015	9/64"	.031	.031	50	.05	.06	.07	.09	.11	.13	.15 0°	.18	.21	.24	.29	.34	.41	.48
0° SOLID STREAM	1/4T 00 02	1/4TT 00 02	9/64"	.036	.036	50	.07	.08	.10	.12	.14	.17	.20 0°	.25	.28	.32	.39	.45	.55	.64
	1/4T 00 03	1/4TT 00 03	9/64"	.043	.043	50	.10	.12	.15	.18	.21	.26	.30 0°	.37	.42	.47	.58	.67	.82	.96
	1/4T 00 04	1/4TT 00 04	9/64"	.052	.052	50	.14	.17	.20	.25	.28	.35	.40 0°	.49	.57	.63	.78	.90	1.10	1.26
	1/4T 00 06	1/4TT 00 06	9/64"	.062	.062	50	.21	.25	.30	.37	.42	.52	.60 0°	.73	.85	.95	1.16	1.34	1.64	1.90
	1/4T 00 08	1/4TT 00 08	9/64"	.072	.072		.28	.33	.40	.49	.56	.69	.80 0°	.98	1.13	1.26	1.55	1.79	2.20	2.52
	1/4T 00 10	1/4TT 00 10	9/64"	5/64"	5/64"		.35	.43	.50	.61	.70	.86	1.0 0°	1.22	1.41	1.58	1.93	2.24	2.74	3.16

* STRAINERS ARE NOT SUPPLIED WITH ORIFICE TIPS OF THIS CAPACITY.

• ORIFICE TIPS OF THIS CAPACITY SUPPLIED IN BRASS ONLY... ALL OTHER CAPACITIES CAN BE SUPPLIED IN BRASS, STAINLESS STEEL OR MONEL. TABULATION IS BASED ON WATER AT TEMPERATURE OF 70°F.

REDRAWN 5-25-55 REPLACES DWG. OF 10-12-49

PATENT NO. 2,621,078
OTHER PATENTS PENDING

DATA SHEET 2851

TABLE 18. — CHARACTERISTIC FLOW RATES OF SPRAYING SYSTEM

SPRAY ANGLE AT 40 p.s.i.	NOZZLE NO. TYPE T FEMALE CONN.	TYPE TT MALE CONN.	"A"	Approx. Equiv. Orifice Dia.	"E"	MESH OF SCREEN	CAPACITY G.P.M. (Gallons Per Minute) AND SPRAY ANGLE AT p.s.i. (Lbs. per square inch)													
							5	7	10	15	20	30	40	60	80	100	150	200	300	400
•	1/4T 110 01	1/4TT 110 01	13/64	.026	.011	100	.03 42°	.04 55°	.05 69°	.06 87°	.07 94°	.09 102°	.10 110°	.12 117°	.14 121°	.16 122°	.19 123°	.22 124°	.27 125°	.32 126°
•	1/4T 110 015	1/4TT 110 015	13/64	.031	.015	100	.05 45°	.06 60°	.07 72°	.09 89°	.11 97°	.13 104°	.15 110°	.18 117°	.21 121°	.24 122°	.29 123°	.34 124°	.41 125°	.48 126
•	1/4T 110 02	1/4TT 110 02	13/64	.036	.018	50	.07 61°	.08 72°	.10 80°	.12 92°	.14 98°	.17 105°	.20 110°	.25 116°	.28 120°	.32 121°	.39 122°	.45 123°	.55 124°	.64 125
110° •	1/4T 110 03	1/4TT 110 03	13/64	.043	.022	50	.11 78°	.12 83°	.15 88°	.18 95°	.21 99°	.26 106°	.30 110°	.37 116°	.42 120°	.47 121°	.58 122°	.67 123°	.82 124°	.94 125
•	1/4T 110 04	1/4TT 110 04	13/64	.052	.026	50	.14 80°	.17 85°	.20 90°	.25 96°	.28 100°	.35 106°	.40 110°	.49 115°	.57 119°	.63 120°	.78 121°	.90 122°	1.10 123°	1.26 124
•	1/4T 110 06	1/4TT 110 06	7/32	.062	.033	50	.21 82°	.25 87°	.30 91°	.37 97°	.42 101°	.52 107°	.60 110°	.73 114°	.85 117°	.95 119°	1.16 120°	1.34 122°	1.64 123°	1.90 124
•	1/4T 110 08	1/4TT 110 08	17/64	.072	.041	50	.28 84°	.33 89°	.40 93°	.49 98°	.56 102°	.69 107°	.80 110°	.98 114°	1.13 117°	1.26 119°	1.55 120°	1.79 121°	2.20 122°	2.5 123
•	1/4T 95 01	1/4TT 95 01	13/64	.026	.012	100	.03 40°	.04 46°	.05 64°	.06 74°	.07 81°	.09 89°	.10 95°	.12 102°	.14 105°	.16 108°	.19 111°	.22 113°	.27 114°	.32 115
•	1/4T 95 015	1/4TT 95 015	13/64	.031	.016	100	.05 42°	.06 49°	.07 66°	.09 75°	.11 82°	.13 90°	.15 95°	.18 102°	.21 105°	.24 108°	.29 111°	.34 113°	.41 114°	.48 115°
	1/4T 95 02	1/4TT 95 02	13/64	.036	.021	50	.07 47°	.08 56°	.10 67°	.12 76°	.14 82°	.17 90°	.20 95°	.25 101°	.28 105°	.32 107°	.39 110°	.45 113°	.55 114°	.64 115
	1/4T 95 03	1/4TT 95 03	13/64	.043	.025	50	.10 49°	.12 62°	.15 69°	.18 77°	.21 83°	.26 91°	.30 95°	.37 100°	.42 104°	.47 105°	.58 108°	.67 111°	.82 112°	.94 113
95°	1/4T 95 04	1/4TT 95 04	13/64	.052	.027	50	.14 52°	.17 64°	.20 71°	.25 79°	.28 84°	.35 91°	.40 95°	.49 100°	.57 103°	.63 104°	.78 106°	.90 108°	1.10 109°	1.26 110
	1/4T 95 06	1/4TT 95 06	7/32	.062	.036	50	.21 57°	.25 68°	.30 74°	.37 82°	.42 86°	.52 92°	.60 95°	.73 99°	.85 101°	.95 102°	1.16 104°	1.34 106°	1.64 107°	1.90 108
	1/4T 95 08	1/4TT 95 08	17/64	.072	.040	50	.28 62°	.33 66°	.40 76°	.49 83°	.56 87°	.69 92°	.80 95°	.98 98°	1.13 100°	1.26 101°	1.55 103°	1.79 105°	2.20 106°	2.52 107
	1/4T 95 10	1/4TT 95 10	3/32	5/64	.054	*	.35 64°	.43 70°	.50 78°	.61 84°	.70 89°	.86 92°	1.0 95°	1.22 97°	1.41 99°	1.58 101°	1.93 103°	2.24 104°	2.74 105°	3.16 106
	1/4T 95 15	1/4TT 95 15	3/32	3/32	.063	*	.53 66°	.63 72°	.75 79°	.92 85°	1.16 90°	1.30 93°	1.5 95°	1.83 96°	2.12 98°	2.37 100°	2.90 102°	3.35 103°	4.10 104°	4.75 105
	1/4T 95 20	1/4TT 95 20	13/64	7/64	.079	*	.71 67°	.84 73°	1.0 80°	1.23 86°	1.41 90°	1.74 93°	2.0 95°	2.45 96°	2.83 97°	3.16 99°	3.88 101°	4.48 103°	5.48 104°	6.33 105
•	1/4T 80 0050	1/4TT 80 0050	13/64	.018	.010	200				.030 50°	.035 61°	.043 72°	.050 80°	.06 91°	.07 95°	.08 99°	.10 101°	.11 101°	.14 101°	.16 102
•	1/4T 80 0067	1/4TT 80 0067	13/64	.021	.012	100			.03 51°	.04 60°	.05 67°	.06 74°	.067 80°	.08 89°	.09 94°	.11 97°	.13 98°	.15 99°	.18 99°	.22 100
	1/4T 80 01	1/4TT 80 01	13/64	.026	.015	100			.05 52°	.06 62°	.07 68°	.09 75°	.10 80°	.12 85°	.14 89°	.16 90°	.19 91°	.22 92°	.27 93°	.32 94°
	1/4T 80 015	1/4TT 80 015	13/64	.031	.018	100		.06 45°	.07 54°	.09 64°	.11 68°	.13 75°	.15 80°	.18 85°	.21 89°	.24 90°	.29 91°	.34 92°	.41 93°	.48 94°
	1/4T 80 02	1/4TT 80 02	13/64	.036	.023	50	.07 34°	.08 48°	.10 56°	.12 65°	.14 69°	.17 76°	.20 80°	.25 84°	.28 88°	.32 89°	.39 90°	.45 91°	.55 92°	.64 93°
80°	1/4T 80 03	1/4TT 80 03	13/64	.043	.028	50	.10 41°	.13 50°	.15 58°	.18 66°	.21 70°	.26 76°	.30 80°	.37 84°	.42 87°	.47 88°	.58 89°	.67 90°	.82 91°	.94 92°
	1/4T 80 04	1/4TT 80 04	13/64	.052	.031	50	.14 45°	.17 53°	.20 60°	.25 67°	.28 71°	.35 77°	.40 80°	.49 85°	.57 86°	.63 87°	.78 88°	.90 89°	1.10 90°	1.26 91°
	1/4T 80 06	1/4TT 80 06	7/32	.062	.040	50	.21 49°	.25 55°	.30 61°	.37 68°	.42 72°	.52 77°	.60 80°	.73 83°	.85 85°	.95 86°	1.16 87°	1.34 88°	1.64 89°	1.90 90°
	1/4T 80 08	1/4TT 80 08	17/64	.072	.044	50	.28 51°	.38 57°	.40 62°	.49 69°	.56 72°	.69 78°	.80 80°	.98 83°	1.13 84°	1.26 85°	1.55 86°	1.79 87°	2.20 88°	2.52 89°
	1/4T 80 10	1/4TT 80 10	3/32	5/64	.050	*	.35 53°	.43 58°	.50 63°	.61 70°	.70 73°	.86 78°	1.0 80°	1.22 82°	1.41 84°	1.58 85°	1.93 86°	2.24 87°	2.74 88°	3.16 89°
	1/4T 80 15	1/4TT 80 15	3/32	3/32	.062	*	.53 55°	.68 60°	.75 64°	.92 71°	1.16 74°	1.30 79°	1.5 80°	1.83 82°	2.12 83°	2.37 84°	2.90 85°	3.35 86°	4.10 87°	4.75 88°
	1/4T 80 20	1/4TT 80 20	13/64	7/64	.075	*	.71 55°	.84 60°	1.0 65°	1.22 71°	1.41 74°	1.73 79°	2.0 80°	2.45 82°	2.82 83°	3.16 84°	3.87 85°	4.47 86°	5.48 87°	6.32 88°
•	1/4T 73 0023	1/4TT 73 0023	13/64	.012	.007	200				.014 42°	.016 50°	.020 64°	.023 73°	.028 83°	.032 89°	.036 92°	.045 95°	.051 97°	.063 98°	.073 99°
•	1/4T 73 0039	1/4TT 73 0039	13/64	.016	.009	200			.020 31°	.024 44°	.028 53°	.034 65°	.039 73°	.048 82°	.055 87°	.062 90°	.075 92°	.087 93°	.11 94°	.12 95°
•	1/4T 73 0077	1/4TT 73 0077	13/64	.022	.013	100		.032 29°	.039 33°	.047 45°	.055 53°	.067 65°	.077 73°	.09 81°	.11 86°	.12 89°	.15 91°	.17 92°	.21 93°	.24 94°
	1/4T 73 0116	1/4TT 73 0116	13/64	.028	.016	100	.04 28°	.05 31°	.06 35°	.07 46°	.08 54°	.10 66°	.116 73°	.14 81°	.16 85°	.18 87°	.22 89°	.26 90°	.32 91°	.37 92°
73°	1/4T 73 0154	1/4TT 73 0154	13/64	.032	.020	100	.05 29°	.06 32°	.08 38°	.09 47°	.11 55°	.13 67°	.154 73°	.19 80°	.22 84°	.24 86°	.30 87°	.34 88°	.42 89°	.48 90°
	1/4T 73 0231	1/4TT 73 0231	13/64	.040	.026	50	.08 30°	.10 33°	.12 39°	.14 49°	.16 56°	.20 67°	.231 73°	.28 79°	.33 83°	.37 85°	.45 86°	.52 87°	.63 88°	.73 89°
	1/4T 73 0308	1/4TT 73 0308	13/64	.045	.031	50	.11 31°	.13 35°	.15 43°	.19 52°	.22 58°	.27 68°	.308 73°	.38 78°	.44 82°	.49 84°	.60 85°	.69 86°	.84 87°	.97 88°
	1/4T 73 0385	1/4TT 73 0385	13/64	.051	.034	50	.14 32°	.16 38°	.19 45°	.24 53°	.27 59°	.33 68°	.385 73°	.47 78°	.54 81°	.61 83°	.74 84°	.86 85°	1.05 86°	1.22 87°
	1/4T 73 0462	1/4TT 73 0462	13/64	.056	.039	50	.16 35°	.19 40°	.23 47°	.28 55°	.33 60°	.40 69°	.462 73°	.57 77°	.65 80°	.73 81°	.90 83°	1.03 84°	1.26 85°	1.46 86°
	1/4T 73 0616	1/4TT 73 0616	13/64	.065	.046	50	.22 39°	.26 44°	.31 50°	.38 58°	.44 63°	.53 70°	.616 73°	.75 77°	.87 79°	.98 80°	1.20 82°	1.38 83°	1.69 84°	1.95 85°
	1/4T 73 0770	1/4TT 73 0770	13/64	.072	.051	50	.27 42°	.32 46°	.38 52°	.47 59°	.54 64°	.67 70°	.770 73°	.94 76°	1.09 77°	1.22 79°	1.49 81°	1.72 82°	2.11 83°	2.44 84°
	1/4T 73 0924	1/4TT 73 0924	7/32	.078	.060	50	.33 44°	.39 49°	.46 55°	.57 61°	.65 65°	.80 70°	.924 73°	1.13 76°	1.30 77°	1.46 78°	1.79 79°	2.06 80°	2.53 81°	2.92 82°

O. HYDRAULIC NOZZLES OPERATED WITH WATER AT GIVEN PRESSURES

SPRAY ANGLE AT 40 p.s.i.	NOZZLE NO. TYPE T FEMALE CONN.	NOZZLE NO. TYPE TT MALE CONN.	"A"	Approx. Equiv. Orifice Dia.	"E"	MESH OF SCREEN	CAPACITY G.P.M. (Gallons Per Minute) AND SPRAY ANGLE AT p.s.i. (Lbs. per square inch) 5	7	10	15	20	30	40	60	80	100	150	200	300	400
65°	• 1/4T 65 0017	1/4TT 65 0017	13/64"	.011	.006	200					.012 44°	.015 58°	.017 65°	.021 73°	.024 77°	.027 81°	.033 84°	.038 86°	.047 90°	.054 94°
	• 1/4T 65 0025	1/4TT 65 0025	13/64"	.013	.007	200					.018 45°	.022 58°	.025 65°	.031 73°	.035 77°	.040 80°	.048 83°	.056 84°	.069 88°	.079 91°
	• 1/4T 65 0033	1/4TT 65 0033	13/64"	.015	.009	200				.021 35°	.023 47°	.029 58°	.033 65°	.040 72°	.047 76°	.052 79°	.064 82°	.074 83°	.090 86°	.11 88°
	• 1/4T 65 0050	1/4TT 65 0050	13/64"	.018	.011	200				.030 40°	.035 48°	.040 58°	.050 65°	.060 72°	.07 75°	.08 78°	.10 81°	.11 82°	.14 84°	.16 86°
	• 1/4T 65 0067	1/4TT 65 0067	13/64"	.021	.013	100		.02 25°	.03 30°	.04 44°	.05 50°	.06 59°	.067 65°	.08 72°	.09 75°	.11 77°	.13 80°	.15 81°	.18 82°	.22 83°
	1/4T 65 01	1/4TT 65 01	13/64"	.026	.017	100		.04 27°	.05 33°	.06 45°	.07 51°	.09 59°	.10 65°	.12 71°	.14 74°	.16 76°	.19 79°	.22 80°	.27 81°	.32 82°
	1/4T 65 015	1/4TT 65 015	13/64"	.031	.023	100		.06 29°	.07 36°	.09 46°	.11 51°	.13 59°	.15 65°	.18 71°	.21 74°	.24 76°	.29 79°	.34 80°	.41 81°	.48 82°
	1/4T 65 02	1/4TT 65 02	13/64"	.036	.024	50	.07 25°	.08 32°	.10 38°	.12 47°	.14 52°	.17 60°	.20 65°	.25 70°	.28 73°	.32 75°	.39 78°	.45 79°	.55 80°	.64 81°
	1/4T 65 03	1/4TT 65 03	13/64"	.043	.027	50	.10 27°	.12 35°	.15 41°	.18 48°	.21 53°	.26 60°	.30 65°	.37 69°	.42 72°	.47 74°	.58 77°	.67 78°	.82 79°	.94 80°
	1/4T 65 04	1/4TT 65 04	13/64"	.052	.034	50	.14 29°	.17 38°	.20 42°	.25 49°	.28 53°	.35 60°	.40 65°	.49 69°	.57 72°	.63 74°	.78 75°	.90 76°	1.10 77°	1.26 78°
	1/4T 65 06	1/4TT 65 06	7/32"	.062	.044	50	.21 32°	.25 39°	.30 43°	.37 50°	.42 54°	.52 61°	.60 65°	.73 69°	.85 72°	.95 73°	1.15 74°	1.34 75°	1.64 76°	1.90 77°
	1/4T 65 08	1/4TT 65 08	17/64"	.072	.050	50	.28 35°	.33 40°	.40 44°	.49 51°	.56 55°	.69 61°	.80 65°	.98 68°	1.13 71°	1.26 72°	1.55 73°	1.79 74°	2.20 75°	2.52 76°
	1/4T 65 10	1/4TT 65 10	3/32"	5/64"	.053	*	.35 38°	.43 42°	.50 46°	.61 52°	.70 56°	.86 61°	1.0 65°	1.22 68°	1.41 71°	1.58 72°	1.93 73°	2.24 74°	2.74 75°	3.16 76°
	1/4T 65 15	1/4TT 65 15	3/32"	3/32"	.070	*	.53 41°	.63 44°	.75 48°	.92 53°	1.16 56°	1.30 62°	1.5 65°	1.83 68°	2.12 70°	2.37 71°	2.90 72°	3.35 73°	4.10 74°	4.75 75°
	1/4T 65 20	1/4TT 65 20	13/64"	7/64"	.076	*	.71 44°	.84 47°	1.0 50°	1.23 54°	1.41 57°	1.74 62°	2.0 65°	2.45 68°	2.83 70°	3.16 71°	3.88 72°	4.48 73°	5.48 74°	6.33 75°

FLAT SPRAY PATTERN

PATENT NO. 2,621,078; OTHER PATENTS PENDING

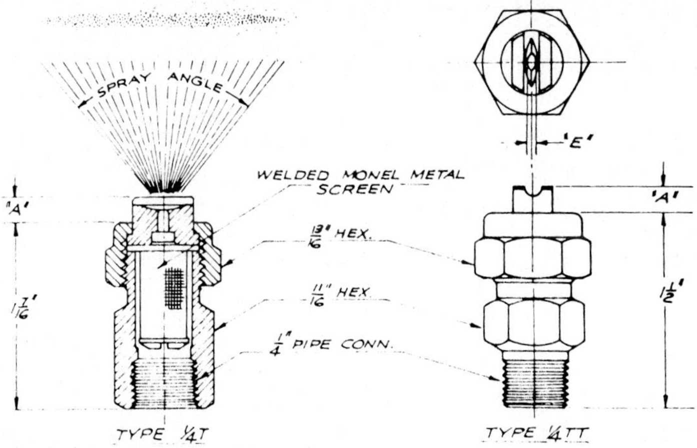

TYPE 1/4T TYPE 1/4TT

* STRAINERS ARE NOT FURNISHED WITH ORIFICE TIPS OF THIS CAPACITY.

• ORIFICE TIPS OF THIS CAPACITY SUPPLIED IN BRASS ONLY... ALL OTHER CAPACITIES CAN BE SUPPLIED IN BRASS, STAINLESS STEEL OR MONEL.

FOR OTHER CAPACITIES AND OTHER PIPE CONNECTIONS SEE DWGS 2882 & 3421

DESCRIPTION	SPRAYING SYSTEMS CO. Engineers and Manufacturers 3201 RANDOLPH ST. BELLWOOD, ILL. SUBURB OF CHICAGO	
TABULATION OF CAPACITIES AND SPRAY ANGLES FOR 1/4T---- AND 1/4TT---- TEEJET NOZZLES	DATE 5-25-55	DWG. NO. 2851

TABLE 19. — CHARACTERISTIC FLOW RATES OF SPRAYING SYSTE

NOZZLE NO.		ORIFICE NO.	CORE NO.	ORIFICE DIA. Inches	CAPACITY G.P.M. (Gallons Per Minute) AND SPRAY ANGLE AT p.s.i. (Lbs. per sq. inch)											
TYPE T FEMALE CONN.	TYPE TT MALE CONN.				10	15	20	25	30	40	60	80	100	150	200	300
1/8 T D2-13	1/8 TT D2-13	D2	13	.041		.058 41°	.064 49°	.070 55°	.075 60°	.08 67°	.10 71°	.11 72°	.12 74°	.14 74°	.16 75°	.18 75°
1/4 T D2-13	1/4 TT D2-13															
1/8 T D3-13	1/8 TT D3-13	D3	13	.047		.064 45°	.071 53°	.077 59°	.08 64°	.09 70°	.11 73°	.12 75°	.13 77°	.16 77°	.18 78°	.20 78°
1/4 T D3-13	1/4 TT D3-13															
1/8 T D4-13	1/8 TT D4-13	D4	13	.063	.070 52°	.082 64°	.09 69°	.10 74°	.11 77°	.12 79°	.14 81°	.16 83°	.17 84°	.20 84°	.23 85°	.27 85°
1/4 T D4-13	1/4 TT D4-13															
1/8 T D2-23	1/8 TT D2-23	D2	23	.041		.070 43°	.078 51°	.085 55°	.092 59°	.10 63°	.13 68°	.14 70°	.16 71°	.19 72°	.21 72°	.25 72°
1/4 T D2-23	1/4 TT D2-23															
1/8 T D3-23	1/8 TT D3-23	D3	23	.047	.065 39°	.078 56°	.087 58°	.096 62°	.10 65°	.12 69°	.14 73°	.16 75°	.18 76°	.21 77°	.24 77°	.28 77°
1/4 T D3-23	1/4 TT D3-23															
1/8 T D4-23	1/8 TT D4-23	D4	23	.063	.082 54°	.100 62°	.113 68°	.13 74°	.14 77°	.15 82°	.19 85°	.21 87°	.23 88°	.28 88°	.32 86°	.38 88°
1/4 T D4-23	1/4 TT D4-23															
1/8 T D5-23	1/8 TT D5-23	D5	23	.078	.095 62°	.117 73°	.13 79°	.15 83°	.16 85°	.18 89°	.22 92°	.25 94°	.28 95°	.34 96°	.38 96°	.46 95°
1/4 T D5-23	1/4 TT D5-23															
1/8 T D6-23	1/8 TT D6-23	D6	23	.094	.112 70°	.136 79°	.15 84°	.17 87°	.19 89°	.21 93°	.26 96°	.29 98°	.32 99°	.39 100°	.45 100°	.54 99°
1/4 T D6-23	1/4 TT D6-23															
1/8 T D2-25	1/8 TT D2-25	D2	25	.041		.10 32°	.12 39°	.13 43°	.14 46°	.16 51°	.19 56°	.22 58°	.25 60°	.29 61°	.34 61°	.41 61°
1/4 T D2-25	1/4 TT D2-25															
1/8 T D3-25	1/8 TT D3-25	D3	25	.047	.10 41°	.12 47°	.14 52°	.15 55°	.17 58°	.19 61°	.23 65°	.26 67°	.29 68°	.35 69°	.40 69°	.48 69°
1/4 T D3-25	1/4 TT D3-25															
1/8 T D4-25	1/8 TT D4-25	D4	25	.063	.15 58°	.18 63°	.21 67°	.23 70°	.25 72°	.29 74°	.35 78°	.40 80°	.45 81°	.54 82°	.62 82°	.75 82°
1/4 T D4-25	1/4 TT D4-25															
1/8 T D5-25	1/8 TT D5-25	D5	25	.078	.18 65°	.22 70°	.25 73°	.28 75°	.30 77°	.35 79°	.42 83°	.48 84°	.54 85°	.65 85°	.75 84°	.90 84°
1/4 T D5-25	1/4 TT D5-25															
1/8 T D6-25	1/8 TT D6-25	D6	25	.094	.23 73°	.28 77°	.32 79°	.35 81°	.39 83°	.44 85°	.54 88°	.62 89°	.70 89°	.85 89°	.97 89°	1.19 88°
1/4 T D6-25	1/4 TT D6-25															
1/8 T D7-25	1/8 TT D7-25	D7	25	.109	.26 79°	.32 83°	.37 85°	.41 87°	.45 89°	.52 91°	.63 92°	.73 93°	.81 93°	.98 92°	1.18 92°	1.37 91°
1/4 T D7-25	1/4 TT D7-25															
1/8 T D8-25	1/8 TT D8-25	D8	25	.125	.31 85°	.38 89°	.43 91°	.48 93°	.53 94°	.61 96°	.75 97°	.89 97°	.97 97°	1.19 96°	1.36 96°	1.68 95°
1/4 T D8-25	1/4 TT D8-25															
1/8 T D10-25	1/8 TT D10-25	D10	25	.156	.38 90°	.46 94°	.54 97°	.60 98°	.65 100°	.76 102°	.93 103°	1.07 103°	1.21 103°	1.48 102°	1.71 102°	2.1 101°
1/4 T D10-25	1/4 TT D10-25															
1/8 T D12-25	1/8 TT D12-25	D12	25	.188	.46 96°	.57 101°	.61 103°	.73 105°	.80 107°	.93 109°	1.15 111°	1.32 112°	1.47 112°	1.81 111°	2.09 111°	2.55
1/4 T D12-25	1/4 TT D12-25															
1/8 T D14-25	1/8 TT D14-25	D14	25	.219	.51 102°	.62 105°	.72 108°	.81 110°	.88 111°	1.03 113°	1.26 114°	1.47 114°	1.65 114°	2.02 113°	2.34 113°	2.89 112°
1/4 T D14-25	1/4 TT D14-25															
1/8 T D2-45	1/8 TT D2-45	D2	45	.041		.13 26°	.14 32°	.16 36°	.18 40°	.20 46°	.25 52°	.28 55°	.32 57°	.38 58°	.44 58°	.53 58°
1/4 T D2-45	1/4 TT D2-45															
1/8 T D3-45	1/8 TT D3-45	D3	45	.047		.14 34°	.17 40°	.18 45°	.20 48°	.23 53°	.28 57°	.33 60°	.36 61°	.44 62°	.51 62°	.62 62°
1/4 T D3-45	1/4 TT D3-45															
1/8 T D4-45	1/8 TT D4-45	D4	45	.063	.18 52°	.22 59°	.25 62°	.28 65°	.31 67°	.36 69°	.43 71°	.50 72°	.56 73°	.68 73°	.78 73°	.95 72°
1/4 T D4-45	1/4 TT D4-45															
1/8 T D5-45	1/8 TT D5-45	D5	45	.078	.23 57°	.28 63°	.32 67°	.36 70°	.39 71°	.45 73°	.55 75°	.64 76°	.71 76°	.86 76°	.99 76°	1.22 75°
1/4 T D5-45	1/4 TT D5-45															
1/8 T D6-45	1/8 TT D6-45	D6	45	.094	.29 65°	.35 70°	.41 73°	.45 75°	.50 77°	.58 79°	.72 80°	.83 81°	.93 81°	1.15 80°	1.33 80°	1.64 79°
1/4 T D6-45	1/4 TT D6-45															
1/8 T D7-45	1/8 TT D7-45	D7	45	.109	.33 73°	.41 78°	.48 81°	.54 83°	.59 84°	.68 86°	.84 87°	.97 87°	1.11 87°	1.35 86°	1.57 86°	1.94 85°
1/4 T D7-45	1/4 TT D7-45															
1/8 T D8-45	1/8 TT D8-45	D8	45	.125	.41 80°	.51 84°	.59 86°	.66 88°	.72 89°	.84 90°	1.04 90°	1.21 90°	1.35 90°	1.68 89°	1.94 89°	2.40 88°
1/4 T D8-45	1/4 TT D8-45															
1/8 T D10-45	1/8 TT D10-45	D10	45	.156	.54 86°	.66 88°	.77 90°	.86 91°	.94 92°	1.10 93°	1.35 93°	1.57 93°	1.77 93°	2.18 92°	2.50 92°	3.10 91°
1/4 T D10-45	1/4 TT D10-45															
1/8 T D12-45	1/8 TT D12-45	D12	45	.188	.67 93°	.82 95°	.95 97°	1.07 98°	1.17 99°	1.36 100°	1.68 101°	1.95 102°	2.20 102°	2.69 101°	3.11 101°	3.80 100°
1/4 T D12-45	1/4 TT D12-45															
1/8 T D14-45	1/8 TT D14-45	D14	45	.219	.75 96°	.92 99°	1.07 101°	1.20 102°	1.32 103°	1.53 104°	1.89 105°	2.19 105°	2.45 105°	3.00 104°	3.49 104°	4.30 103°
1/4 T D14-45	1/4 TT D14-45															
1/8 T D16-45	1/8 TT D16-45	D16	45	.250	.86 103°	1.08 108°	1.25 108°	1.40 109°	1.54 110°	1.79 111°	2.20 112°	2.57 112°	2.89 112°	3.54 111°	4.11 111°	5.20 110°
1/4 T D16-45	1/4 TT D16-45															

HYDRAULIC NOZZLES OPERATED WITH WATER AT GIVEN PRESSURES

NOZZLE NO.		ORIFICE NO.	CORE NO.	ORIFICE DIA. Inches	CAPACITY G.P.M. (Gallons Per Minute) AND SPRAY ANGLE AT p.s.i. (Lbs. per sq. inch)											
TYPE T FEMALE CONN.	TYPE TT MALE CONN				10	15	20	25	30	40	60	80	100	150	200	300	400
1/8T D2-46 1/4T D2-46	1/8TT D2-46 1/4TT D2-46	D2	46	.041				.22	.24 14°	.27 15°	.33 18°	.37 20°	.42 21°	.50 21°	.57 20°	.68 19°	.78 18°
1/8T D3-46 1/4T D3-46	1/8TT D3-46 1/4TT D3-46	D3	46	.047			.23 14°	.25 16°	.28 18°	.32 20°	.39 23°	.45 24°	.51 24°	.61 23°	.70 22°	.86 21°	.99 21°
1/8T D4-46 1/4T D4-46	1/8TT D4-46 1/4TT D4-46	D4	46	.063	.28 16°	.34 20°	.39 23°	.44 25°	.48 26°	.56 29°	.68 32°	.78 33°	.88 33°	1.07 32°	1.23 32°	1.52 31°	1.76 31°
1/8T D5-46 1/4T D5-46	1/8TT D5-46 1/4TT D5-46	D5	46	.078	.38 22°	.47 28°	.54 33°	.60 35°	.66 37°	.77 39°	.94 41°	1.10 42°	1.25 42°	1.50 41°	1.73 41°	2.13 40°	2.47 40°
1/8T D6-46 1/4T D6-46	1/8TT D6-46 1/4TT D6-46	D6	46	.094	.55 31°	.67 38°	.78 42°	.87 45°	.95 47°	1.10 48°	1.35 49°	1.58 50°	1.73 50°	2.16 49°	2.50 48°	3.06 47°	3.52 47°
1/8T D7-46 1/4T D7-46	1/8TT D7-46 1/4TT D7-46	D7	46	.109				.98 48°	1.10 50°	1.22 51°	1.39 53°	1.72 55°	1.97 56°	2.22 56°	2.73 55°	3.15 54°	3.85 53°	4.42 53°
1/8T D8-46 1/4T D8-46	1/8TT D8-46 1/4TT D8-46	D8	46	.125						1.45 57°	1.59 58°	1.84 60°	2.25 61°	2.62 62°	2.93 62°	3.60 61°	4.17 60°	5.05 59°	5.88 58°
1/8T D10-46 1/4T D10-46	1/8TT D10-46 1/4TT D10-46	D10	46	.156							2.15 65°	2.48 66°	3.05 67°	3.53 68°	3.96 68°	4.83 66°	5.59 65°	6.80 64°	7.90 64°

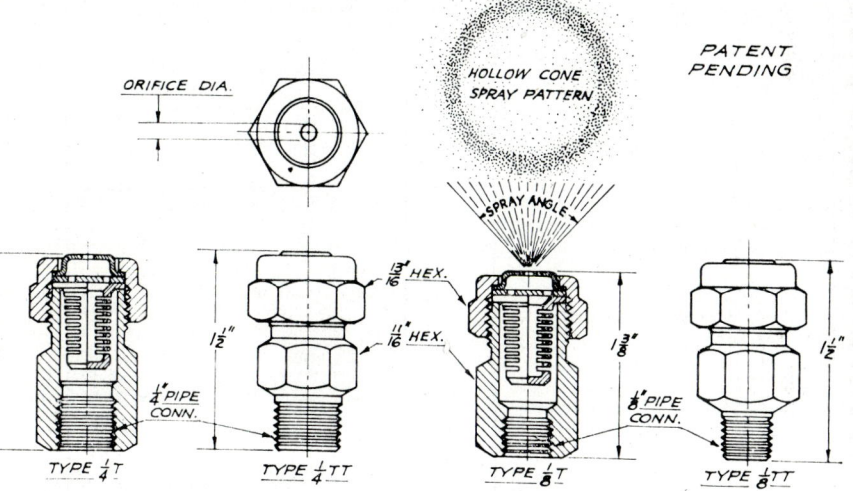

- TABULATION IS BASED ON WATER AT TEMPERATURE OF 70°F.
- NOZZLE CAPS AND BODIES ARE THE SAME AS USED ON OTHER TEEJET NOZZLES. THEREFORE, REGULAR TEEJET TIPS AND STRAINERS ARE INTERCHANGEABLE WITH THOSE OF THE DISC TYPE.
- THE DISC ORIFICES ARE INTERCHANGEABLE, AS ARE THE CORES.
- DISC ORIFICE IS MADE OF HARDENED STAINLESS STEEL. ALL OTHER PARTS ARE MADE OF BRASS.
- STRAINER IS APPROXIMATELY EQUIVALENT TO 25 MESH SCREEN.

SUPERSEDES DWG. 4498 OF 10-14-49

DESCRIPTION: DISC TYPE TEEJET NOZZLES

SPRAYING SYSTEMS CO.
Engineers and Manufacturers
3201 RANDOLPH ST.
BELLWOOD, ILL.
SUBURB OF CHICAGO

DATE: 5-26-50
DWG. NO. 4498

TABLE 20(a). — ACRES COVERED FOR GIVEN FIELD LENGTHS AND SWATH WIDTHS

Field length	Swath width (ft)																
	20	25	30	35	40	45	50	55	65	75	85	95	100
ft (mi)	.... acres* ....
1 320 (1/4)	0.6	0.75	0.9	1.1	1.2	1.4	1.5	1.7	2.0	2.3	2.5	2.9	3.0
2 640 (1/2)	1.2	1.5	1.8	2.1	2.4	2.7	3.0	3.3	3.9	4.5	5.1	5.7	6.1
3 960 (3/4)	1.8	2.3	2.7	3.2	3.6	4.1	4.6	5.1	5.9	6.8	7.7	8.7	9.1
5 280 (1)	2.4	3.05	3.6	4.2	4.8	5.5	6.1	6.7	7.8	9.1	10.2	11.6	12.1
2	4.9	6.05	7.2	8.4	9.8	10.9	12.1	13.3	15.6	18.2	20.8	23.0	24.2
3	7.25	9.1	10.8	12.6	14.5	16.4	18.2	20.0	23.4	27.3	30.8	34.6	36.4
4	9.7	12.1	14.4	16.8	19.4	21.8	24.2	26.6	31.2	36.4	41.2	46.0	48.5
5	12.1	15.15	18.0	21.0	24.2	27.3	30.3	33.3	39.3	45.5	51.4	57.6	60.6

* Acres = (Length in ft × width in ft)/43 560.

TABLE 20(b). — HECTARES COVERED FOR GIVEN FIELD LENGTHS AND SWATH WIDTHS

Field length	Swath width (m)												
	7.5	10	15	20	25	30	35	40	45	50	55	65
m (km)	.... hectares* ....
250 (1/4)	0.19	0.25	0.38	0.5	0.63	0.75	0.88	1.0	1.13	1.25	1.38	1.63
500 (1/2)	0.38	0.5	0.75	1.0	1.25	1.5	1.75	2.0	2.25	2.5	2.75	3.25
750 (3/4)	0.56	0.75	1.13	1.5	1.9	2.25	2.63	3.0	3.38	3.75	4.13	4.88
1 000 (1)	0.75	1.0	1.5	2.0	2.5	3.0	3.5	4.0	4.5	5.0	5.5	6.5
2	1.5	2.0	3.0	4.0	5.0	6.0	7.0	8.0	9.9	10.0	11.0	13.0
3	2.3	3.0	4.5	6.0	7.5	9.0	10.5	12.0	13.5	15.0	16.5	19.5
4	3.0	4.0	6.0	8.0	10.0	12.0	14.0	16.0	18.0	20.0	22.0	26.0
5	3.75	5.0	7.5	10.0	12.5	15.0	17.5	20.0	22.5	25.0	27.5	32.5

* Hectares = (Length in m × width in m)/10 000.

TABLE 21(a). — HECTARE PER MINUTE COVERAGE FOR GIVEN VELOCITIES AND SWATH WIDTHS

Velocity	Swath width (m)										
	7.5	10.	12.5	15.	17.5	20.	22.5	27.5	32.5	37.5	42.5	47.5
km/hr					hectares/minute*						
100	1.3	1.7	2.1	2.5	2.9	3.3	3.8	4.8	5.4	6.3	7.0	7.9
110	1.4	1.8	2.3	2.8	3.2	3.7	4.1	5.0	6.0	6.9	7.8	8.7
120	1.5	2.0	2.5	3.0	3.5	4.0	4.5	5.5	6.5	7.5	8.5	9.5
130	1.6	2.2	2.7	3.2	3.8	4.3	4.9	6.0	7.0	8.1	9.2	10.3
140	1.8	2.3	2.9	3.5	4.1	4.7	5.3	6.4	7.6	8.8	10.0	11.1
150	1.9	2.5	3.1	3.8	4.4	5.0	5.6	6.9	8.1	9.4	10.6	11.9
160	2.0	2.7	3.3	4.0	4.7	5.3	6.0	7.3	8.7	10.0	11.3	12.7
170	2.1	2.8	3.5	4.2	4.9	5.7	6.4	7.8	9.2	10.6	12.0	13.4
180	2.3	3.0	3.8	4.5	5.3	6.0	6.8	8.3	9.8	11.3	12.8	14.3
190	2.4	3.2	4.0	4.8	5.5	6.3	7.1	8.7	10.3	11.9	13.5	15.0

* Hectares/minute = (width in m × km/hr)/600.

TABLE 21(b). — ACRE PER MINUTE COVERAGE FOR GIVEN VELOCITIES AND SWATH WIDTHS

Velocity	Swath width (ft)										
	20	25	30	35	40	45	55	65	75	85	95	100
mi/hr					acres/minute*						
60	2.4	3.0	3.6	4.2	4.8	5.4	6.6	7.8	9.0	10.2	11.4	12.2
65	2.6	3.3	3.9	4.5	5.2	5.8	7.2	8.4	9.7	11.0	12.4	13.0
70	2.8	3.5	4.2	4.9	5.7	6.3	7.7	9.1	10.5	12.0	13.4	14.0
75	3.0	3.8	4.5	5.2	6.0	6.7	8.3	9.7	11.2	12.7	14.3	15.0
80	3.2	4.0	4.8	5.6	6.4	7.2	8.8	10.4	12.0	13.6	15.2	16.0
85	3.4	4.3	5.1	5.9	6.8	7.6	9.3	11.0	12.7	14.4	16.1	17.0
90	3.6	4.5	5.4	6.3	7.2	8.1	9.9	11.7	13.5	15.3	17.1	18.0
95	3.8	4.8	5.7	6.6	7.6	8.5	10.4	12.3	14.2	16.1	18.0	19.0
100	4.0	5.0	6.0	7.0	8.0	9.0	11.0	13.0	15.0	17.0	19.0	20.0
110	4.4	5.5	6.6	7.7	8.8	9.9	12.1	14.3	16.5	18.7	20.9	22.0

* Acres/minute = (width in ft × mi/hr)/495.

TABLE 22(a). — REQUIRED JET DIAMETERS FOR GIVEN TANK SIZES AND PRESSURES

Tank diameter		Pressure		Jet diameter for each 30 cm (1 ft) of tank length	
cm	(ft)	kg/cm²	(lbf/in²)	cm	(in)
61	(2)	1.8	(25)	0.32	(1/8)
		5.3	(75)	0.24	(3/32)
64	(2.1)	1.8	(25)	0.48	(3/16)
		5.3	(75)	0.32	(1/8)
76	(2.5)	1.8	(25)	0.56	(7/32)
		5.3	(75)	0.40	(5/32)

TABLE 22(b). — FLOW RATES (WATER) FOR GIVEN JET DIAMETERS AND PRESSURES

Jet diameter		Pressure: kg/cm² (lbf/in²)		
		1.8 (25)	3.5 (50)	5.3 (75)
cm	(in)	l/min (gals/min)	l/min (gals/min)	l/min (gals/min)
0.16	(1/16)	2.8 (0.75)	4.7 (1.25)	5.6 (1.5)
0.24	(3/32)	4.7 (1.25)	6.6 (1.75)	8.5 (2.25)
0.32	(1/8)	8.5 (2.25)	11.4 (3.0)	14.5 (3.75)
0.40	(5/32)	13.3 (3.5)	17.0 (4.5)	22.7 (6.9)
0.48	(3/16)	19.0 (5.0)	27.0 (7.0)	30.0 (8.0)
0.56	(7/32)	27.0 (7.0)	36.0 (9.5)	45.0 (12.0)
0.64	(1/4)	34.0 (9.0)	45.0 (12.0)	57.0 (15.0)

TABLE 23(a). — TOXICITY RATINGS OF CHEMICALS USED IN PEST CONTROL

Do not depend exclusively on toxicity values as the *only* factor to be considered regarding the toxic effect of a chemical on humans or other animals. Users of pesticides should be concerned with the hazard(s) associated with exposure to the chemical and not exclusively the *toxicity* of the material itself. These two terms are *not* synonymous.

Toxicity is the inherent capacity of a substance to produce injury or death.

Hazard is a function of two primary variables, *toxicity* and *exposure*, and is the probability that injury will result from the use of a substance in a given formulation, quantity, or manner. Some hazards are not toxic to humans or other animals. For example, sulfur, oils and numerous other chemicals are considered safe or relatively safe to animals, but may pose a considerable hazard to some plants (phytotoxicity).

A compound may be extremely toxic, but present little hazard to the applicator or others when used:

(*a*) in a very dilute formulation;

(*b*) in a formulation that is not readily absorbed through the skin or readily inhaled;

(*c*) only occasionally and under conditions to which humans are not exposed;

(*d*) only by experienced applicators who are properly equipped to handle the chemical safely.

On the other hand, a chemical may exhibit a relatively low mammalian *toxicity*, but still present a *hazard* because it is normally used in concentrated form, which may readily be absorbed or inhaled; or it may be used frequently by nonprofessionals (e.g., home gardeners) who are not aware of the possible hazards to which they are being exposed.

Human experience with exposure to chemicals is the best guide to human toxicity. No actual scientific tests have been conducted in which humans have been subjected to lethal doses of chemicals. However, the effects of some chemicals on humans have been obtained from reports of accidental exposures or suicides. This information is at best usually incomplete or fragmentary, and we must therefore rely upon data obtained from animal tests. The toxicity values obtained from these tests may vary according to the species of test animal used, the test method used, sex of the species, whether the animals have been fasted or not, the state of their health, the purity of the chemical tested, the medium in which the toxicant is administered, the route of administration, and the length of time and frequency of exposure. Therefore, these data cannot be interpreted directly in terms of human toxicity, and extrapolation of them as to their effect upon humans must be pursued with caution.

Toxicity values are expressed as *acute oral* LD_{50} in terms of milligrams of the substance per kilogram (mg/kg) of body weight of the test animal; as *acute dermal* LD_{50} in terms of mg/kg; or as *inhalation* data LC_{50} in terms of micrograms of mist or dust per litre of air ($\mu g/l$) or parts per million by volume of gas or vapour (ppm). One microgram (μg) equals one millionth of a gram.

TABLE 23(b). — COMBINED TABULATION OF TOXICITY CLASSES

Toxicity rating	Toxicity expressed in common terms	Routes of absorption		
		LD_{50} Single oral dose for rats (mg/kg)	LD_{50} Single dermal dose for rabbits (mg/kg)	Probable lethal oral dose for man
1	Extremely toxic	1 or less	20 or less	A taste, a grain
2	Highly toxic	1-50	20-200	A pinch, 1 teaspoon
3	Moderately toxic	50-500	200-1 000	1 teaspoonful to 2 tablespoons
4	Slightly toxic	500-5 000	1 000-2 000	1 ounce to 1 pint
5	Practically non-toxic	5 000-15 000	2 000-20 000	1 pint to 1 quart
6	Relatively harmless	15 000	20 000	1 quart

MODIFIED FROM: M.N. Gleason, R.E. Gosselin, and H.C. Hodge, *Clinical Toxicology o Commercial Products*, 2nd ed. The Williams and Wilkins Company, Baltimore, Md. (U.S.A.), 1963; Wayland J. Hayes, Jr., M.D., Ph.D. (U.S. Department of Health, Education and Welfare, Public Health Service, Communicable Disease Center, Toxicology Section, Atlanta, Georgia), *Clinical Handbook on Economic Poisons*, 1963.

TABLE 24. — ESTIMATED RELATIVE ACUTE TOXIC HAZARDS OF PESTICIDES TO SPRAYMEN.* The estimates of hazards are based primarily on the observed acute dermal and, to a lesser extent, oral toxicity of these compounds in experimental animals. Use experience has also been considered It should be noted that the classification into toxicity groups is both approximate and relative. These toxicity categories are not related to specific categories spelled out for label requirements.

Most dangerous	Dangerous	Less dangerous**	Least dangerous
Carbanolate (Tenrik)	Aldrin	Akton	Abate
Carbofuran (Furadan)	Azodrin	Azinophos-Methyl (Guthion)	Alar
Dasanit	Baygon (propoxur)	Binapacryl (Morocide)	Aramite
Demeton (Systox)	Bidrin (dicrotophos)	BHC	Bromophos
Disulfoton (Di-System)	Bux	Chlordane	Captan
Dyfonate	Carbophenothion (Trithion)	Ciodrin	Carbaryl (Sevin)
Lannate (methomyl)	Dichlorvos (DDVP)	Coumaphos (Co-Ral)	Chlorobenzilate
Mevinphos (Phosdrin)	Dieldrin	Diazinon	2,4-D
Monitor	Dioxathion (Delnav)	Dicapthon	DDT
Parathion	DNOC	Dichloroethyl ether	Dicofol (Kelthane)
Phorate (Thimet)	DNOSBP	Dimethoate	Dilan
Schradan (OMPA)	Endrin	Dinobuton (Dessin)	Dinocap (Karathane)
TEPP	EPN	Endosulfan (Thiodan)	Diquat
Zinophos (Cynem)	Methyl parathion	Ethion	Gardona (Rabon)
	Nicotine	Fenthion (Baytex)	Malathion
	Paraquat	Fundal (chlorphenamidine)	Maneb
	Pentachlorophenol	Galecron (chlorphenamidine)	Methoxychlor
	Phosphamidon	Heptachlor	Mirex
	Zectram	Imidan	Morestan
	Zolone	Lead arsenate	NAA
		Lindane	Omite
		Naled (Dibrom)	Perthane
		Oxydemetonmethyl	Phostex
		(Meta-Systox-R)	Piperonyl butoxide
		Ruelene (crufomate)	Ronnel (Korlan)
		Toxaphene	Rotenone
		Trichlorfon (Dipterex)	Sulphenone
		VC-13	TDE (DDD)
		Vapam	Tedradifon (Tedion)

* Modified from Wolf & Durham, 1966, WSU (Eastern Washington Fertilizer and Pesticide Conference).
** The fumigant compounds acrylonitrile, D-D, or Vidden D, and Telone have toxicities which would indicate their placement in the "less dangerous" category; however, special note should be taken of the fact that the volatility of these compounds and their capacity to produce irritation of skin, eyes, and other tissues indicate that appropriate caution should be exercised in their use.

TABLE 25. — CONVERSION TABLES FOR METRIC AND U.K. AND U.S. WEIGHTS AND MEASURES

Linear measure

Metric				English, U.S.A.				
km	×	1 000	= m	mi	×	175	=	yds
m	×	100	= cm	yds	×	3	=	ft
cm	×	10	= mm	rods	×	16.5	=	ft
mm	×	1 000	= μm (microns, micrometres)	mi	×	5 280	=	ft
				ft	×	12	=	in
m	×	10^6	= μm (microns, micrometres)	nautical mi	×	1.16	=	statute mi

Metric - English				English - Metric				
km	×	0.6213	= mi (statute)	mi	×	1.6093	=	km
m	×	1.0936	= yds	yds	×	0.9144	=	m
m	×	3.28	= ft	ft	×	0.3048	=	m
cm	×	0.3937	= in	in	×	2.54	=	cm
μm	×	0.000393	= in	in	×	25 400	=	μm

Square measure

Metric				English, U.S.A.				
km²	×	10^2	= ha	acres	×	43 560	=	ft²
are	×	10^2	= m²	acres	×	4 840	=	yds²
ha	×	10^4	= m²	mi²	×	640	=	acres
ha	×	10^2	= are	ft²	×	144	=	in²
cm²	×	10^4	= m²					
mm²	×	10^2	= cm²					

Metric - English				English - Metric				
				mi²	×	259	=	ha
ha	×	107 641	= ft²	mi²	×	2.59	=	km²
ha	×	2.471	= acres	acres	×	0.4047	=	ha
km²	×	0.3861	= mi²	ft²	×	0.0929	=	m²
m²	×	10.764	= ft²	in²	×	6.452	=	cm²
cm²	×	0.155	= in²					

Cubic measure (Volume)

Metric				English, U.S.A.				
m³	×	999.97	= l (litres)	yds³	×	27	=	ft³
m³	×	10^6	= cm³ (cc)	ft³	×	1 728	=	in³
cm³	×	10^3	= mm³	gal	×	4	=	qts
cm³	×	10^{12}	= μm³ (cubic microns)	qt	×	2	=	pts
l (litres)	×	10^2	= cl (centilitres)	gal (Imp.)	×	1.201	=	gal (U.S.A.)
l	×	10^3	= ml (millilitres)	ft³	×	7.48	=	gal (U.S.A.)
ml	×	1.000028	= cm³					

English, U.S.A. (cont.)

ft³	×	6.25	=	gal (Imp).
gal (U.S.A.)	×	231	=	in³
gal (Imp.)	×	276.5	=	in³
fl oz (Imp.)	×	0.9607	=	fl oz (U.S.A.)
fl oz (U.S.A.)	×	1.805	=	in³
fl oz (Imp.)	×	1.734	=	in³
gal (U.S.A.)	×	128	=	fl oz (U.S.A.)
gal (Imp.)	×	160	=	fl oz (Imp.)
pt (U.S.A.)	×	16	=	fl oz (U.S.A.)
pt (Imp.)	×	20	=	fl oz (Imp.)

Metric - English, U.S.A.

l (litres)	×	0.264	=	gal (U.S.A.)
l	×	1.057	=	qt (U.S.A.)
l	×	0.88	=	qt (Imp.)
l	×	2.11	=	pts (U.S.A.)
ml	×	0.0338	=	fl oz (U.S.A.)
ml	×	0.0352	=	fl oz (Imp.)
l	×	61.025	=	in³
l	×	0.0353	=	ft³
cm³	×	0.061	=	in³
m³	×	35.314	=	ft³
m³	×	1.308	=	yds³

English, U.S.A. - Metric

gal (U.S.A.)	×	3.785	=	l (litres)
gal (Imp.)	×	4.544	=	l
qt (U.S.A.)	×	0.946	=	l
pt (Imp.)	×	0.568	=	l
fl oz (U.S.A)	×	29.6	=	ml
fl oz (Imp.)	×	28.4	=	ml
in³	×	16.38	=	cm³
ft³	×	28.33	=	l
ft³	×	0.0283	=	m³
yd³	×	0.764	=	m³

Velocity

Metric

km/hr	×	16.67	=	m/min
km/hr	×	0.278	=	m/sec
km/hr	×	27.78	=	cm/sec

Metric - English, U.S.A

km/hr	×	54.68	=	ft/min
km/hr	×	0.9113	=	ft/sec
km/hr	×	0.6214	=	mi/hr
m/min	×	3.281	=	ft/min

English, U.S.A.

mi/hr	×	88	=	ft/min
mi/hr	×	1.467	=	ft/sec
knot	×	1.689	=	ft/sec

English, U.S.A. - Metric

mi/hr	×	1.609	=	km/hr
mi/hr	×	26.82	=	m/min
ft/min	×	0.3048	=	m/min
ft/sec	×	30.48	=	cm/sec

Pressure

kg/m²	×	10^4	=	kg/cm²
lb/in²	×	144	=	lb/ft²

Metric - English, U.S.A.

kg/m²	×	0.205	=	lb/ft²
kg/cm²	×	14.22	=	lb/in²

English, U.S.A. - Metric

lb/ft²	×	4.88	=	kg/m²
lb/in²	×	0.0703	=	kg/cm²

Density of water

Metric

litre	×	1	=	kg
cm³	×	0.9999	=	g
ml	×	1	=	g

English, U.S.A.

pt (Imp.)	×	1.25	=	lb
pt (U.S.A.)	×	1	=	lb
gal (Imp.)	×	10	=	lb
gal (U.S.A.)	×	8.32	=	lb
ft³	×	62.37	=	lb

Metric - English, U.S.A.

g/ml	×	8.347	=	lb/gal
kg/m³	×	0.0624	=	lb/ft³

English, U.S.A. - Metric

lb/gal	×	0.1198	=	g/ml
lb/ft³	×	16.1	=	kg/m³

Application conversions

SOLIDS

Metric - English, U.S.A.

g/ha	×	0.0142	=	oz/acre
kg/ha	×	14.27	=	oz/acre
kg/ha	×	0.894	=	lb/acre

English, U.S.A. - Metric

oz/acre	×	70	=	g/ha
oz/acre	×	0.07	=	kg/ha
lb/acre	×	1.12	=	kg/ha

LIQUIDS

Metric - English, U.S.A.

ml/ha	×	0.1369	=	fl oz/acre (U.S.A.)
l/ha	×	8.557	=	pt/acre (U.S.A.)
ml/ha	×	0.144	=	fl oz/acre (Imp.)
l/ha	×	7.125	=	pt/acre (Imp.)
l/ha	×	0.107	=	gal/acre (U.S.A.)
l/ha	×	0.09	=	gal/acre (Imp.)

English, U.S.A. - Metric

fl oz/acre (U.S.A.)	×	73	=	ml/ha
pt/acre (U.S.A.)	×	1.71	=	l/ha
fl oz/acre (Imp.))	×	87.7	=	ml/ha
pt/acre (Imp.)	×	1.4	=	l/ha
gal/acre (U.S.A.)	×	9.35	=	l/ha
gal/acre (Imp.)	×	11.23	=	l/ha

WEIGHT AND MASS

Metric

g	×	10^{12}	=	picograms
g	×	10^9	=	nanograms
g	×	10^6	=	micrograms
kg	×	10^3	=	g
tonne	×	10^3	=	kg

English, U.S.A.

lb		× 16	=	oz (avoirdupois)
cwt (Eng., long)	×	112	=	lbs
cwt (U.S.A., short)	×	100	=	lbs
ton	×	20	=	cwt
ton (Eng., long)	×	2 240	=	lbs
ton (U.S.A., short)	×	2 000	=	lbs

Metric - English, U.S.A.

g	×	0.0353	=	oz (avoirdupois)
g	×	453.6	=	lb
kg	×	2.205	=	lb
ton	×	2 205	=	lb
ton	×	0.984	=	ton (Eng., long)
ton	×	1.102	=	ton (U.S.A., short)

English, U.S.A. - Metric

oz (avoirdupois)	×	28.35	=	g
lb	×	0.453	=	kg
ton (Eng., long)	×	1 016	=	kg
ton (U.S.A., short)	×	907	=	kg

Other measurements

lb/gal (U.S.A.)	× 12	per cent*	
lb/gal (Imp.)	× 10	per cent*	
g/l	× 0.1	per cent*	
mg/kg		ppm	
lb/10^6 lb		ppm	

lb/ton (short)	× 500	=	ppm
oz/gal(water, U.S.A.)	× 0.781	=	per cent
per cent	× 10^4	=	ppm
(C^0 × 9/5) + 32		=	degrees F
(F^0 − 32) × 5/9		=	degrees C

* By weight in water.

BIBLIOGRAPHICAL REFERENCES

AKESSON, N.B. & YATES, W.E. Research and development of chemical distribution
1963 equipment for agricultural aircraft in California. *J. R. aeronaut. Soc.*, 67(636): 760-767.

AKESSON, N.B. & YATES, W.E. Airplane application of bulk fertilizer. *Trans. Am.*
1964 *Soc. agric. Engrs*, 7(2): 137-140, 141.

AKESSON, N.B. & YATES, W.E. Problems relating to the application of agricultural
1964 chemicals and resulting drift residues. *A. Rev. Ent.*, 9: 285-318.

AKESSON, N.B., YATES, W.E. & BURGOYNE, W.E. *Use of helicopters for applying chem-*
1966 *ical spray materials*. St. Joseph, Mich., American Society of Agricultural Engineers. ASAE Paper 66-104.

AKESSON, N.B., YATES, W.E. & WILCE, S.E. Performance of atomizers for aircraft
1969 chemical application. *Proc. 4th int. agric. Aviat. Congr.*, Kingston, Ontario, p. 254-264.

AKESSON, N.B., WILCE, STEPHEN E. & YATES, W.E. *Pesticide chemicals as environmental*
1970 *contaminants*. St. Joseph, Mich., American Society of Agricultural Engineers. ASAE Paper 70-101.

AKESSON, N.B., WILCE, S.E. & YATES, W.E. Confining aerial applications to treated
1971 fields, a realistic goal. *Agrichemical Age*.

ALEXANDER, GRAHAM & TULLETT, J.S. The super men. In *Agricultural aviation in New*
1967 *Zealand*, Chapter 2. Wellington, Reed.

AMSDEN, R.C. *Aircraft applications*. Chesterford Park Research Station Report,
1962a Fisons.

AMSDEN, R.C. Reducing the evaporation of sprays. *Agric. Aviat.*, 4(3): 89-93
1962b

AMSDEN, R.C. Factors affecting the application of low volume sprays. *J.R. aeronaut.*
1964 *Soc.*, 68: 535-539.

ARMSTRONG, J.A. & RANDALL, A.P. Determination of spray distribution patterns in
1969 forest application. *Proc. 4th int. agric. Aviat. Congr.*, Kingston, Ontario, p. 196-204.

ASSOCIATE COMMITTEE AGR. AND FORESTRY AVIATION NRC OF CANADA. *Recommended*
1968 *certification for qualification of pilots and ground personnel for agricultural and forestry aviation*. Montreal. Document 7192, part 19, AFA-TN4.

AZAR'YAN, M.B. et al. *The application of aviation in agriculture and forestry*. Washing-
1966 ton, D.C., U.S. Department of Commerce. Translation FTD-MT 24-101-68. (*Tr. Izdatel'stvo Transport*. Moskva. 381 p.)

BAILEY, J. BLAIR & SWIFT, JOHN E. *Pesticides information and safety manual*. Berkeley,
1968 Calif., University of California, Agricultural Extension Service.

BALS, E.J. Rotary atomization. *Agric. Aviat.*, 12(3): 85-90.
1970

BALTIN, F. & BRANDT, R. An improvement of the method of measuring the distri-
1966 bution of fertilizers applied from the air. *Proc. 3rd int. agric. Aviat. Congr.*,
Arnhem, p. 119-128.

BARNES, J.M. *Toxic hazards of certain pesticides to man.* Geneva, World Health Or-
1953 ganization. WHO Monograph Series No. 16.

BARUCH, D. Some medical aspects in agricultural flights relating to fatigue among agri-
1970 cultural pilots. *Aerospace Med.*, 41(4): 447-450.

BATES, E.N. California rice land seeded by airplane. '*Agric. Engng*, St. Joseph, Mich.,
1930 11(2); 69-70.

BAUER, SEPP. Use of the helicopter against weaver bird in the Sudan. *Agric. Aviat.*,
1966 8(3): 90-95.

BERNER, W.H. Aerial application accidents. A critical review. *Aerial Applicator*,
1970 8(4): 6-7.

BOVING, P.A., STEVENS, L.E. & WINTERFIELD, R.G. A new hydraulic drive system for
1971 the Piper Pawnee. *Aerial Applicator*, 9(6): 4,5.

BRAZELTON, R.W. *New concepts in aircraft granular applicators.* St. Joseph, Mich.,
1970 American Society of Agricultural Engineers, ASAE Paper 70-657.

BRAZELTON, R.W. et al. Distribution of dry materials (aircraft). *Proc 4th int. agric.*
1969 *Aviat. Congr.*, Kingston, Ontario.

BROOKS, F.A. The drifting of poisonous dusts applied by airplanes and ground rigs.
1947 *Agric. Engng*, St. Joseph, Mich., 28(6): 233-234.

BROWN, A.W.A. *Insect control by chemicals.* London, Chapman Hall.
1951

BRUGGINK, GERARD M., BARNES, ALFRED C., JR. & GREGG, LEE W. Injury reduc-
1964 tion trends in agricultural aviation. *Aerospace Med.*, 35(5): 472-475.

BURNETT, G.F. Research in east Africa on the control of tsetse flies from the air. *Agric.*
1962 *Aviat.*, 4(3): 79-87.

BUTLER, B.J., AKESSON, N.B. & YATES, W.E. Use of spray adjuvants to reduce drift.
1969 *Trans. Am. Soc. agric. Engrs*, 12(2): 182-186.

BYASS, J.B. & COURSHEE, R.J. *A study of the methods of measuring small spray drops.*
1951 Silsoe, National Institute of Agricultural Engineering. NIAE Report No. 31.

CALIFORNIA (STATE). DEPARTMENT OF AGRICULTURE. *Injurious materials.* Sacramento,
1927 Calif. California Admin. Code, Title 3, Agriculture Section 2461.

CIBA. *The principles of waterless spraying.* MacQuaig, R.D. Locust control and the
1969 development of ULV spraying. Hadaway, A.B. & Johnstone, D.R., Developments in low volume spraying. Basel, CIBA Agro Chemicals. Technical Monograph No. 2.

COAD, B., JOHNSON, R.E. & MCNEIL, G.L. *Dusting cotton from airplanes.* Washington,
1922 D.C., U.S. Department of Agriculture. Bulletin No. 1204.

COUTTS, H.H. Preliminary tests with the UCAR nozzle. *Agric. Aviat.*, 9(4): 123-124.
1967

COUTTS, H.H. & YATES, W.E. Analysis of spray droplet distribution from agricultural
1968   aircraft. *Trans. Am. Soc. agric. Engrs*, 11(1): 25-27.

Crop dusting: legal problems in a new industry. *Stanford Law Rev.* 6(1): 95.
1953

CUMMINS, T.D. Chairman, Yolo Co. (Calif.) Board of Supervisors. Ordinance 201.
1946

CUTKAMP, I.K., HESS, A.D. & KEENER, G.G. Factors influencing spray and aerosol ap-
1950   plications by airplane. *J. econ. Ent.*, 43(4): 456-462.

DAVID, W.A.L. *et al.* Factors influencing the interaction of insecticidal mists and flying
1946   insects. Part I. Design of a spray testing chamber and some of its properties.
Bull. ent. Res., 36:373 394, 1946. Part II. The production and behavior of
kerosene base insecticidal spray mists and their relation to flying insects. *Bull.
ent. Res.*, 3 v., 37: 1-27, 1946. Part III. Biological factors. *Bull. ent. Res.*,
3 v., 37: 177-190, 1946. Part IV. Some experiments with adjuvants. *Bull.
ent. Res.* 3 v., 37: 393-398, 1946.

DAVIS, J.M. A rapid method for estimating aerial spray deposits. *J. econ. Ent.*, 46(4):
1953   696-698.

FAO. *Guidelines for legislation concerning the registration for sale and marketing of pesti-
1969   cides*. Rome, FAO/World Health Organization.

FINKELSTEIN, HAROLD, ed. *Air pollution aspects of pesticides*. Springfield, Va., U.S.
1969   Department of Commerce, National Technical Information Service. National
Air Pollution Control Administration PB 188-091.

FRAZER, R.P. The fluid kinetics of application of pesticidal chemicals. In *Advances
1958   in pest-control research*. New York, Wiley.

FRENCH, O.C. The use of airplanes for agricultural pest control. *Agric. Engng*, St.
1947   Joseph, Mich., 28(6): 240, 242, 244.

GALIPAULT, J.B. Agricultural flight safety: the procedure turn-around. *Agric. Aviat.*,
1966   8(1): 14-18.

GIBSON, E.A. Aircraft in agriculture. *J. R. aeronaut. Soc.*, 62: 423-436.
1958

GIESEKE, JAMES A. & MITCHELL, RALPH I. Size measurement of collected drops. *J. chem.
1965   Engng. Data*, 10(4): 350-353.

GILLIES, P.A., WOMELDORF, D.J. & WALSH, J.D. A bioassay method for measuring
1968   mosquito larvicide depositions. *Mosquito News*, 28(3): 415-421.

GINSBERG, J.M. *Proceedings of the 18th annual meeting of the New Jersey Mosquito
1931   Extermination Association*.

GLANCY, MICHAEL *et al.* Low volume application of the insecticides for the control
1966   of adult mosquitoes. *Mosquito News*, 26(3): 356-359.

GOODHUE, LYLE D. Aerosols and their application. *J. econ. Ent.*, 39(4): 506-509.
1949

GREEN, H.L. & LANE, W.R. *Particulate clouds, dusts, smokes, and mists*, p. 415-416.
1964   London, Spon.

HADAWAY, A.B. *et al*. *The effect of particle size in the contact toxicity of insecticides to
1970   adult mosquitoes*. Geneva, World Health Organization. WHO/VBC/70-213.

Handbook for agricultural pilots. The Hague, International Agricultural Aviation
1968 Centre.

HIMEL, C.M. The fluorescent particle spray droplet tracer method. *J. econ. Ent.*,
1969a 62(4): 912-916.

HIMEL, C.M. The optimum size for insecticide droplets. *J. econ. Ent.*, 62(4): 920-925.
1969b

HIMEL, C.M. Spray droplet size in the control of spruce budworm, bollweevil, bollworm
1969c and cabbage. *J. econ. Ent.*, 62(4): 916-918.

HOCKING, K.S., YEO, D. & AINSTAY, D.G. Aircraft applications of insecticides in east
1954 Africa. No.6. Applications of a coarse aerosol containing DDT to control the
tsetse flies *Glossina morsitans* (West.), *Glossina swynmertoni* (Aust.) and *Glossina pallidepes* (Aust.). *Bull. ent. Res.*, 45(3): 585-603.

HOFFMAN, ROBERT, A. & LUNDQUIST, ARTHUR W. Fumigation properties of several new
1949 insecticides. *J. econ. Ent.*, 42(3): 436-438.

HOUSER, J.S. The airplane in Catalpa sphinx control. *Ohio agric. exp. Station Bull.*,
1922 7(7-8): 126-136.

HOWARD, W.E. *et al.* Range rodent control by plane. *Calif. Agric.*, 10(10): 8,9.
1956

HUSMAN, C.N. *et al.* Equipment for the dispersal of DDT insecticides by means of aircraft.
1947 Washington, D.C., U.S. Department of Agriculture. A.R.S. Ent. and P.W.ET. 228.

INTERNATIONAL CIVIL AVIATION ORGANIZATION. *Safety in aerial work.* Part 1. Agri-
1971 cultural operations. Montreal. ICAO circular 85 AN/71.

JANSSEN, M.R. *Self analysis of the aerial applicator's business.* Third Agricultural
1959 Aviation Conference, Milwaukee, Wisc. Washington, D.C., U.S. Department
of Agriculture, Agricultural Research Service.

JOHNSTONE, D.R. Some current ideas regarding the rotating distributor. *J.R. acronaut.*
1963 *Soc.*, 67(636): 767-770.

JOHNSTONE, D.R. Formulations and atomization. *Proc. 4th int. agric. Aviat. Congr.*,
1969 Kingston, Ontario, p. 225-233.

JOHNSTONE, H.F., ed. *Handbook on aerosols.* Washington, D.C., Atomic Energy Com-
1950 mission.

JOHNSTONE, H.F., WINSCHE, W.E. & SMITH, L.W. The dispersion and deposition of
1949 aerosols. *Chem. Rev.*, 44(2): 353-371.

JOSE, M.J. How much formal training? *Proc. 4th int. agric. Aviat. Congr., Kingston,*
1969 *Ontario,* p. 400-404.

JOYCE, R.J.V. *et al.* Waterless spraying in East Pakistan using the Decca navigation
1968 system. *Agric. Aviat.*, 10(4): 118-124, 128-136.

KEELER, A.A. A note on the development of power line markers for aerial crop spraying
1971 operations in Australia. *Agric. Aviat.*, 13(3): 81-86.

KILGORE, W.W., YATES, W.E. & OGAWA, J.M. Evaluation of concentrate and dilute
1964 ground air-carrier and aircraft spray coverages. *Hilgardia*, 35(19).

KILPATRICK, JOHN W., TONN, ROBERT J. & JATANASEN, SUJARTI. Evaluation of ultra-
1970 low-volume insecticide dispersing systems for use in single-engine aircraft and

their effectiveness against *Aedes aegypti* population in southeast Asia. *World Health Organization Bull.*, 42:1-14.

KING, W.V. & BRADLEY, G.H. Airplane dusting in the control of malaria mosquitoes.
1926 Washington, D.C., U.S. Department of Agriculture. Circular No. 367.

KRUSE, C.W., HESS, A.D. & LUDVIK, G.F. The performance of liquid spray nozzles
1949 for aircraft insecticide operations. *J. natn. Malar. Soc.*, 312-334.

LAMER, V.K. & HOCKBERG, SEYMORE. The laws of deposition and the effectiveness of
1949 insecticidal aerosols. *Chem. Rev.*, 44(2): 341-352.

LANGE, B. Aerial control of continental voles (*Microtus Arvalis Pallar*) in Germany.
1960 *Agric. Aviat.*, 2(2): 43-48.

LATTA, RANDALL *et al.* The effect of particle size and velocity of movement of DDT
1947 aerosols in a wind tunnel on the mortality of mosquitoes. *J. Wash. Acad. Sci.*, 37(1): 397-407.

LEE, C.W. *et al.* Modifications to micronair equipment and assessment for fine aerosol
1969 emission in tsetse fly control. *Agric. Aviat.*, 11(1): 12-17.

LEE K.C. & STEPHENSON, J. The distribution of solid materials. *Proc. 4th int. agric.*
1969 *Aviat. Congr.*, Kingston, Ontario, p. 203-213.

LOFGREN, C.S., MOUNT, G.S. & FORD, H.R. A fuselage-mounted boom and spray
1970 system for drift-spraying of insecticide for mosquito control. *Agric. Aviat.*, 12(4): 120-124.

LOMAS, J., FRANKEL, H. & HIRSCH, I. Meteorological considerations in determining the
1954 permissible time for cotton spraying from the air in Israel. *Agric. Met.* Amsterdam, 1(3): 225-234.

MAAN, W.J. Editorial. Fifty years of agricultural aviation. *Agric. Aviat.*, 3(3): 77-80.
1961

MAAN, W.J. *The use of aircraft in the mechanization of agricultural production.* Rome,
1965 FAO. Informal Working Bulletin No. 26.

MAAN, W.J. Editorial. IAAC at work. *Agric. Aviat.*, 9(4): 111-113.
1967

MAAS, W. *ULV application and formulation techniques.* Amsterdam, N.V. Phillips-
1971 Duphar Crop Prof. Div.

MACQUAIG, R.D. The collection of spray droplets by flying locusts. *Bull. ent. Res.*,
1962 53, Part 1: 111-123.

MAY, K.R. An improved spinning top homogeneous spray apparatus. *J. appl. Phys.*,
1949 20(10): 932-938.

MAY, K.R. The measurement of airborne droplets by the magnesium oxide method.
1950 *J. scient. Instrum.*, 27: 128-130.

MCGOVERN, E.G., FALES, J.H. & GOODHUE, L.D. Testing aerosols against houseflies.
1943 *Soap sanit. Chem.*, 19(9): 99-107.

MCMAHON, J.M. & WOOLEY, D.H. Requirements for agricultural aircraft pilots view-
1963 point and engineering aspects. *J.R. aeronaut. Soc.*, 67 (636): 770-774.

MESSENGER, KENNETH. Use of C-47 airplane for baiting and spraying. *Agric. Chem.*,
1953 15.

MESSENGER, KENNETH. Agricultural aviation in the USA. *Agric. Aviat.*, 2(1): 8-11.
1960

MILLER, A.W.D. & CHADWICK, P.R. Swath marking in aerial spraying. *Agric. Aviat.*,
1963   5(4): 114-120.

MOUNT, G.A. *et al.* A new ultra-low volume cold aerosol nozzle for dispersal of insec-
1970   ticides against adult mosquitoes. *Mosquito News*, 30(1): 56-59.

MOUNT, G.A. Optimum droplet size for adult mosquito control with space sprays or
1970   aerosols of insecticides. *Mosquito News*, 39(1): 70-75.

MULLER, PAUL. The development of an aerial jeep to be used as an agricultural work
1969   horse. *Proc. 4th int. agric. Aviat. Congr., Kingston, Ontario*, p. 108-115.

NATIONAL AERIAL APPLICATORS SAFETY COMMISSION. Watch those turns. *Aerial Ap-
1970   plicator*, 8(2): 12-13.

NEILLIE, C.R. & HOUSER, J.L. Fighting insects with airplanes. *Natn. geogr. Mag.*,
1922   41(3): 232-238.

NORMAN, N.D. Economic factors affecting agricultural aircraft operations. *Report of
1959   the 1st International Aviation Congress, Cranfield, England, 1959*, p. 286-297.

OGAWA, J.M., YATES, W.E. & KILGORE, W.W. Susceptibility of almond leaf to coryneum
1964   blight, and evaluation of helicopter spray applications for disease control.
Hilgardia, 35(19).

ORR, CLYDE. *Particulate technology.* New York, Macmillan.
1966

PAGET-CLARK, C.D. Electronic guidance systems used in fire-ant eradication programs.
1971   *Agric. Aviat.*, 13(3): 89-90.

PARAMONOV, A. YA. *Aerial control of forest pests in USSR.* Washington, D.C., U.S.
1963   Department of Commerce, Office of Technical Services. (Tr. *Issledovaniya i
Materialy*, Series 1, No. 49, p. 5-76.)

PARKER, J.D., COLLINGS, B.G.P. & KAHUMBURA, J.M. Preliminary tests of a suction
1971   spray nozzle for use with aircraft spraying systems. *Agric. Aviat.*, 13(1): 24-28.

PASQUILL, F. *Atmospheric diffusion.* London, Van Nostrand.
1962

PHILPOTTS, L.E. & RIECKON, T.O. Costs of operating aircraft for agricultural and for-
1969   estry use in 1967. *Canadian Farm Econ.*, 4(2): 24-29.

PORCH, HARRIETT E. Aircraft in agriculture and forestry of communist China. *Agric.
1967   Aviat.*, 9(3): 80-84.

POTTS, SAMUEL F. Spraying woodlands with an autogyro for control of the gypsy moth.
1929   *J. econ. Ent.*, 32(3): 381-386.

RATHBURN, CARLISLE B. & MISEROCCHI, ARTURO F. Direct photography of air-borne
1967   droplets. *J. econ. Ent.*, 60(1): 247-254.

RATHBURN, CARLISLE B. *et al.* Evaluation of the ultra-low-volume aerial spray technique
1969   by use of caged adult mosquitoes. *Mosquito News*, 29(3): 376-381.

RAZAK, KENNETH, YATES, W.E. & AKESSON, N.B. Operations analysis and evaluation
1963   of agricultural aircraft. *Trans. Am. Soc. agric. Engrs*, (6)1): 43-54.

REICH, GEORGE A. & BERNER, WILLIAM H. Aerial application accidents 1963-1966.
1968   *Archs envir. Hlth*, 17: 776-784.

RILEY, J.A. & GILES, W.L. Agricultural meteorology in relation to the use of pesticides
1955   in the USA. *Agric. Met.* Amsterdam, 2(4): 225-245.

ROBERTS, SEAN C. & SMITH, MICHAEL R. *The evaluation of a positive energy distribution
1963 system for the aerial application of agricultural materials.* Mississippi State
University, Aero-physics Department. Research Note No. 20.

ROHRMAN, DOUGLAS F. Pesticide laws and legal implications of pesticide use. Parts
1968 I and II. *Fd, Drug, Cosmetic Law J.*, 23(3,4): 142-161, 172-184.

SAYER, H.J. An ultra low volume spraying technique for the control of the desert locust
1959 *Schistocerca gregaria* (Forsk.) *Bull. ent. Res.*, 50:270-386.

SAYER, H.J. *Lectures of training course on aerial spraying.* Rome, FAO. United Na-
1966 tions Development Programme. Progress Report USSF/DL/TC/15.

SAYER, H.J. Ultra-low-volume spraying systems comparison and assessment. *Agric.*
1969 *Aviat.*, 11(3):78-85.

SCHULTZ, H.B., AKESSON, N.B. & YATES, W.E. The delayed sea breezes in the Sacra-
1961 mento valley and resulting favorable conditions for applications of pesticides.
Bull. Am. met. Soc., 42 (10): 679-687.

SCOVILLE, HERBERT, JR. *Insect control: the development of equipment for the dispersal
1946 of DDT.* Washington, D.C., National Research Council. NRC Insect Control
Committee Report No. 180; Coordination Center Review No. 8.

SEBORA, L.H. *et al.* Performance of aerial spray equipment used to disperse DDT at
1946 Orlando, Fla. *Mosquito News*, 6(4):169-177.

SHARP, R.B. & BUFTON, L.P. A rapid method for the direct measurement of oil spray
1963 drops. *J. agric. Engng Res.*, 8(3): 237-239.

SHAY, J.R. *Remote sensing with special reference to agriculture and forestry.* Washing-
1970 ton, D.C., National Academy of Sciences.

SLADE, DAVID H., ed. *Meteorology and atomic energy.* Springfield, Va., U.S. Depart-
1968 ment of Commerce, Air Research Laboratories. Chapters 2,3,4, on Atmospheric transport and diffusion; Chapter 6 on Instruments.

SMITH, CHARLES M. & GOODHUE, LYLE D. Particle size in relation to insecticide ef-
1942 ficiency. *Ind. Engng Chem. analyt. Edn*, 34(4); 490-493.

SMITH, M.R. Evaluation of aircraft performance. *Proc. 4th int. agric. Aviat. Congr., King-*
1969 *ston, Ontario*, p. 83-95.

Some agricultural aviation statistics. *Agric. Aviat.*, 11(4): 14.
1969

SPRAYING SYSTEMS CO. (NOZZLE ATOMIZERS). *Drop size distributions from pressure
1968 nozzles.* Wheaton, Ill. U.S.A.

SPUYBROEK, P.H.G. Air worthiness of agricultural airplanes. *Agric. Aviat.*, 1(3):
1969 62-66.

STEEL, F.J. The control and regulation of agricultural aviation in New Zealand. *Proc.
1969 4th int. agric. Aviat. Congr., Kingston, Ontario*, p. 367-373.

SWEENY, T.E. & NIXON, W.B. Air cushion vehicles and their promise. *Mech. Engng,*
1968 1968: 40-43.

TAKANAGA, TAKASHI. *Pest control machinery.* Nissin-Cho, Omiya-shi Saitama, Japan,
1969 Laboratory, Institute of Agricultural Machinery. (Personal communication)

UNDERWOOD, D.J. The contribution of avionics (navigation systems). *Proc. 4th int.
1969 agric. Aviat. Congr., Kingston, Ontario*, p. 121-127.

VAN BEMMEL, P.M. *Crop spraying by air.* Delft, Shell Research Laboratory. Report
1953 1298-M.

VAUGHAN, L.M., MAGILL, P.L. & MCMULLEN, R.W. *Application of the fluorescent par-*
1965 *ticle tracer technique to forest spray research.* Palo Alto, Calif., Metronics Associates, Inc. Mem. Rpt. 278-1.

WATKINS, T.C. & NORTON, L.B. *Handbook of insecticide dust diluents and carriers.*
1955 Rev. by D.E. Weidhaas and J.L. Brown. Caldwell, N.J., Dorland.

WEICK, FRED E. et al. *Handbook on aerial application in agriculture.* College Station.
1956 Texas, Texas Agricultural and Mechanical College, Extension Service.

WEIDHAAS, D.E. et al. Relationship of minimum lethal dose to the optimum size of
1970 droplets of insecticides for mosquito control. *Mosquito News,* 30(2): 195-200.

WHITTAM, DONALD. Aircraft guidance methods for pest control in the U.S. *Agric.*
1962 *Aviat.,* 4(1): 8-15.

WILCE, S.E., et al. Drop size control and aircraft spray equipment. *Agric. Aviat.,* 16(1):
1974 7-16.

WOOLEY, D.H. A note on helicopter spray distribution. *Agric. Aviat.,* 5(2): 43-47.
1963

WORLD HEALTH ORGANIZATION. *Specifications for pesticides used in public health.*
1967 3rd ed. Geneva.

WORLD HEALTH ORGANIZATION. *Control of pesticides: a survey of existing legislation.*
1970 Geneva.

YATES, W.E. Spray pattern analysis and evaluation of deposits from agricultural air-
1970 craft. *Trans. Am. Soc. agric. Engrs,* 5(1): 1-10.

YATES, W.E. & AKESSON, N.B. Fluorescent tracers for quantitative microresidue analysis.
1963 *Trans. Am. Soc. agric. Engrs,* 6(2): 104-107, 114.

YATES, W.E. & AKESSON, N.B. Characteristics of drift deposits resulting from pesticide
1966 applications with aircraft. *Proc. 3rd int. agric. Aviat. Congr., Arnhem,* p. 129-142.

YATES, W.E. & AKESSON, N.B. Selection of spray equipment and operating techniques
1966 for agricultural aircraft. *Proc. 3rd int. agric. Aviat. Congr., Arnhem,* p. 251-277.

YATES, W.E., AKESSON, N.B. & COUTTS, H.H. Drift hazards related to ultra-low volume
1967 and diluted sprays, applied by agricultural aircraft. *Trans. Am. Soc. agric.
Engrs,* 10 (5): 628-632, 638.

YATES, W.E., OGAWA, J.M. & AKESSON, N.B. *Spray distributions in peach orchards from*
1968 *helicopter and ground applications.* American Society of Agricultural Engineers,
1968 winter meeting. Paper No. 68-617.

YEO, D. Droplet size distribution of liquid in a flat spray. *Br. J. appl. Phys.,* (3):
1952 189-192.

YEO, D. Assessment of rotary atomizers fitted to a Cessna (182) aircraft. *Agric. Aviat.,*
1961 3(4): 131-135.

YEO, D. & THOMPSON, B.W. The deposition in open country of a coarse aerosol released
1954 from an aircraft. Aircraft application of insecticide in east Africa. *Bull. ent.
Res.,* 45 (1): 79-92.

YEOMANS, A.H. *Directions for applying windborne aerosols for insect control out of doors.*
1950   Washington, D.C., U.S. Department of Agriculture, Agricultural Research Service. ET-282.

YEOMANS, A.H. & ROGERS, E.E.   Factors influencing deposit of spray droplets. *J. econ.*
1953   *Ent.*, 40(1): 57-60.

YUILL, J.S. & EATON, C.B.   The airplane in forest-pest control. *In* U.S. Department
1949   of Agriculture. *Yearbook of agriculture 1949*, p. 471-476. Washington, D.C.